Dimensions Math®
Teacher's Guide PKB

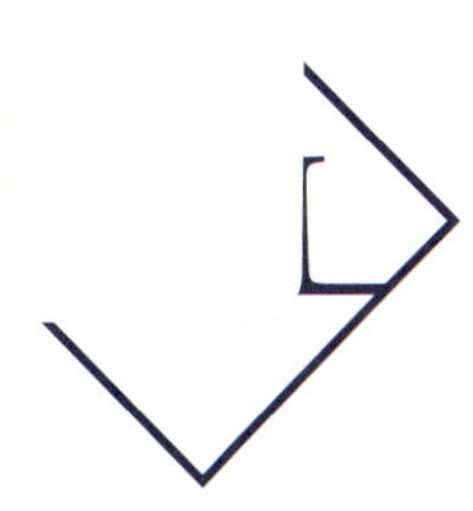

Authors and Reviewers

Tricia Salerno

Pearly Yuen

Jenny Kempe

Dr. Leslie Arceneaux

Allison Coates

Cassandra Turner

Singapore Math Inc.

Published by Singapore Math Inc.

19535 SW 129th Avenue
Tualatin, OR 97062
www.singaporemath.com

Dimensions Math® Teacher's Guide Pre-Kindergarten B
ISBN 978-1-947226-29-6

First published 2018
Reprinted 2019, 2020, 2021

Printed in China

Acknowledgments

Editing by the Singapore Math Inc. team.
Design and illustration by Cameron Wray with Carli Bartlett.

Contents

Chapter | Lesson | Page

Notes

Dimensions Math® Curriculum

The **Dimensions Math®** series is a Pre-Kindergarten to Grade 5 series based on the pedagogy and methodology of math education in Singapore. The main goal of the **Dimensions Math®** series is to help students develop competence and confidence in mathematics.

The series follows the principles outlined in the Singapore Mathematics Framework below.

Pedagogical Approach and Methodology

- Through Concrete-Pictorial-Abstract development, students view the same concepts over time with increasing levels of abstraction.
- Thoughtful sequencing creates a sense of continuity. The content of each grade level builds on that of preceding grade levels. Similarly, lessons build on previous lessons within each grade.
- Group discussion of solution methods encourages expansive thinking.
- Interesting problems and activities provide varied opportunities to explore and apply skills.
- Hands-on tasks and sharing establish a culture of collaboration.
- Extra practice and extension activities encourage students to persevere through challenging problems.
- Variation in pictorial representation (number bonds, bar models, etc.) and concrete representation (straws, linking cubes, base ten blocks, discs, etc.) broaden student understanding.

Each topic is introduced, then thoughtfully developed through the use of a variety of learning experiences, problem solving, student discourse, and opportunities for mastery of skills. This combination of hands-on practice, in-depth exploration of topics, and mathematical variability in teaching methodology allows students to truly master mathematical concepts.

Singapore Mathematics Framework

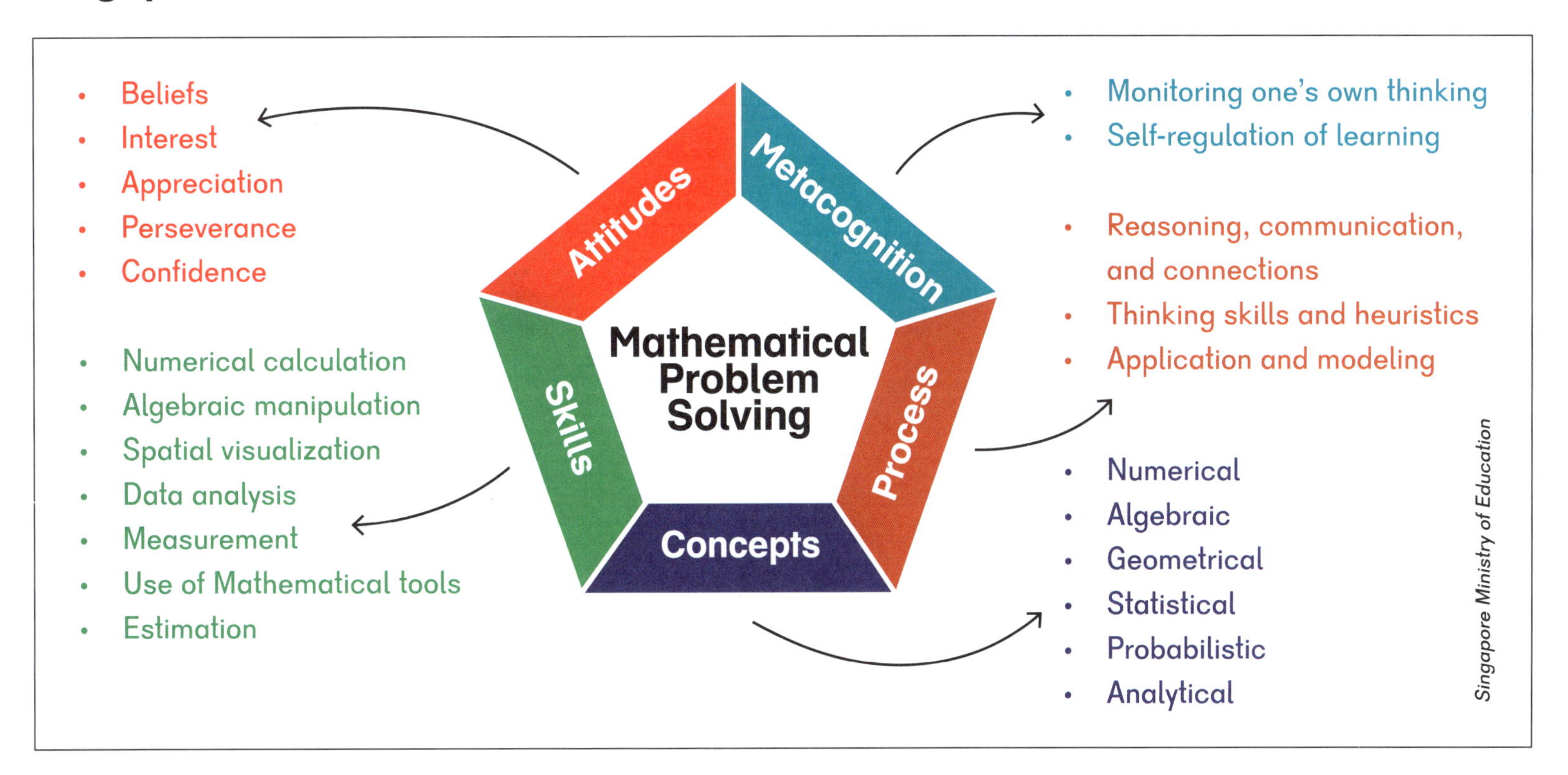

Dimensions Math® Program Materials

Textbooks

Textbooks are designed to help students build a solid foundation in mathematical thinking and efficient problem solving. Careful sequencing of topics, well-chosen problems, and simple graphics foster deep conceptual understanding and confidence. Mental math, problem solving, and correct computation are given balanced attention in all grades. As skills are mastered, students move to increasingly sophisticated concepts within and across grade levels.

Students work through the textbook lessons with the help of five friends: Emma, Alex, Sofia, Dion, and Mei. The characters appear throughout the series and help students develop metacognitive reasoning through questions, hints, and ideas.

A pencil icon at the end of the textbook lessons links to exercises in the workbooks.

Workbooks

Workbooks provide additional problems that range from basic to challenging. These allow students to independently review and practice the skills they have learned.

Teacher's Guides

Teacher's Guides include lesson plans, mathematical background, games, helpful suggestions, and comprehensive resources for daily lessons.

Tests

Tests contain differentiated assessments to systematically evaluate student progress.

Online Resources

The following can be downloaded from dimensionsmath.com.

- **Blackline Masters** used for various hands-on tasks.
- **Letters Home** to be emailed or sent home with students for continued exploration. These outline what the student is learning in math class and offer suggestions for related activities at home. Reinforcement at home supports deep understanding of mathematical concepts.
- **Videos** of popular children songs used for singing activities.
- **Material Lists** for each chapter and lesson, so teachers and classroom helpers can prepare ahead of time.
- **Activities** that can be done with students who need more practice or a greater challenge, organized by concept, chapter, and lesson.
- **Standards Alignments** for various states.

Using the Teacher's Guide

This guide is designed to assist in planning daily lessons, and should be considered a helping hand between the curriculum and the classroom. It provides introductory notes on mathematical content, key points, and suggestions for activities. It also includes ideas for differentiation within each lesson, and answers and solutions to textbook and workbook problems.

Each chapter of the guide begins with the following.

- **Overview**

  Includes objectives, vocabulary, and suggested number of class periods for each chapter.

- **Notes**

  Highlights key learning points, explains the purpose of certain activities, and helps teachers understand the flow of topics throughout the year.

- **Materials**

  Lists materials, manipulatives, and Blackline Masters used in the Explore and Learn sections of the guide. It also includes suggested storybooks and snacks. Blackline Masters can be found at dimensionsmath.com.

The guide goes through the Chapter Openers, Daily Lessons, and Practices of each chapter in the following general format.

- **Explore**

  Introduces students to math concepts through hands-on activities. Depending on the classroom, students may be in a circle for Explore then transition to their tables for Learn.

- **Learn**

  Summarizes the main concepts of the lesson, including exercises from Look and Talk pages of the corresponding textbook pages.

- **Whole Group Play** & **Small Group Center Play**

  Allows students to practice concepts through hands-on tasks and games, including suggestions for outdoor play (most of which can be modified for a gymnasium or classroom).

  - **Activities**

    These recurring activities are for groups of 2 to 4 students at one center. They can be used in math class or at other times of the day.

    Art **Building** **Counting** **Dramatic Play** **Geoboard**
    Sort **Match** **Patterns** **Reading Time** **Music**
    Ten-frame Flash **Roll and Compare** **Number Composition**

- **Extend**

  This expands on **Explore**, **Learn**, and **Play** and provides opportunities for students to deepen their understanding and build confidence.

Discussion is a critical component of each lesson. Have students share their ideas with a partner, small group, or the class as often as possible. As each classroom is different, this guide does not anticipate all situations. Teachers are encouraged to elicit higher level thinking and discussion through questions like these:

- Why? How do you know?
- Can you explain that?
- Can you draw a picture of that?
- Does your answer make sense? How do you know?
- How is this task like the one we did before? How is it different?
- What did you learn before that can help you to solve this problem?
- What is alike and what is different about that?
- Can you solve that a different way?
- How do you know it's true?
- Can you restate or say in your own words what your classmate shared?

Lesson structures and activities do not have to conform exactly to what is shown in the guide. Teachers are encouraged to exercise their discretion in using this material in a way that best suits their classes.

Dimensions Math® Scope & Sequence

PKA

Chapter 1
Match, Sort, and Classify

Red and Blue
Yellow and Green
Color Review
Soft and Hard
Rough, Bumpy, and Smooth
Sticky and Grainy
Size — Part 1
Size — Part 2
Sort Into Two Groups
Practice

Chapter 2
Compare Objects

Big and Small
Long and Short
Tall and Short
Heavy and Light
Practice

Chapter 3
Patterns

Movement Patterns
Sound Patterns
Create Patterns
Practice

Chapter 4
Numbers to 5 — Part 1

Count 1 to 5 — Part 1
Count 1 to 5 — Part 2
Count Back
Count On and Back
Count 1 Object
Count 2 Objects
Count Up to 3 Objects
Count Up to 4 Objects
Count Up to 5 Objects
How Many? — Part 1
How Many? — Part 2
How Many Now? — Part 1
How Many Now? — Part 2
Practice

Chapter 5
Numbers to 5 — Part 2

1, 2, 3
1, 2, 3, 4, 5 — Part 1
1, 2, 3, 4, 5 — Part 2
How Many? — Part 1
How Many? — Part 2
How Many Do You See?
How Many Do You See Now?
Practice

Chapter 6
Numbers to 10 — Part 1

0
Count to 10 — Part 1
Count to 10 — Part 2
Count Back
Order Numbers
Count Up to 6 Objects
Count Up to 7 Objects
Count Up to 8 Objects
Count Up to 9 Objects
Count Up to 10 Objects — Part 1
Count Up to 10 Objects — Part 2
How Many?
Practice

Chapter 7
Numbers to 10 — Part 2

6
7
8
9
10
0 to 10
Count and Match — Part 1
Count and Match — Part 2
Practice

PKB

Chapter 8
Ordinal Numbers

First
Second and Third
Fourth and Fifth
Practice

Chapter 9
Shapes and Solids

Cubes, Cylinders, and Spheres
Cubes
Positions
Build with Solids
Rectangles and Circles
Squares
Triangles

Dimensions Math® Scope & Sequence

Count Up to 10 Things — Part 2
Recognize the Numbers 6 to 10
Write the Numbers 6 and 7
Write the Numbers 8, 9, and 10
Write the Numbers 6 to 10
Count and Write the Numbers 1 to 10
Ordinal Positions
One More Than
Practice

Chapter 4
Shapes and Solids

Curved or Flat
Solid Shapes
Closed Shapes
Rectangles
Squares
Circles and Triangles
Where is It?
Hexagons
Sizes and Shapes
Combine Shapes
Graphs
Practice

Chapter 5
Compare Height, Length, Weight, and Capacity

Comparing Height
Comparing Length
Height and Length — Part 1
Height and Length — Part 2
Weight — Part 1
Weight — Part 2
Weight — Part 3
Capacity — Part 1
Capacity — Part 2
Practice

Chapter 6
Comparing Numbers Within 10

Same and More
More and Fewer
More and Less
Practice — Part 1
Practice — Part 2

KB

Chapter 7
Numbers to 20

Ten and Some More
Count Ten and Some More
Two Ways to Count
Numbers 16 to 20
Number Words 0 to 10
Number Words 11 to 15
Number Words 16 to 20
Number Order
1 More Than or Less Than
Practice — Part 1
Practice — Part 2

Chapter 8
Number Bonds

Putting Numbers Together — Part 1
Putting Numbers Together — Part 2
Parts Making a Whole
Look for a Part
Number Bonds for 2, 3, and 4
Number Bonds for 5
Number Bonds for 6
Number Bonds for 7
Number Bonds for 8
Number Bonds for 9
Number Bonds for 10
Practice — Part 1
Practice — Part 2
Practice — Part 3

Chapter 9
Addition

Introduction to Addition — Part 1
Introduction to Addition — Part 2
Introduction to Addition — Part 3
Addition
Count On — Part 1
Count On — Part 2
Add Up to 3 and 4
Add Up to 5 and 6
Add Up to 7 and 8
Add Up to 9 and 10
Addition Practice
Practice

Chapter 10
Subtraction

Take Away to Subtract — Part 1

Dimensions Math® Scope & Sequence

Compare Numbers to 20
Addition
Subtraction
Practice

Chapter 6
Addition to 20

Add by Making 10 — Part 1
Add by Making 10 — Part 2
Add by Making 10 — Part 3
Addition Facts to 20
Practice

Chapter 7
Subtraction Within 20

Subtract from 10 — Part 1
Subtract from 10 — Part 2
Subtract the Ones First
Word Problems
Subtraction Facts Within 20
Practice

Chapter 8
Shapes

Solid and Flat Shapes
Grouping Shapes
Making Shapes
Practice

Chapter 9
Ordinal Numbers

Naming Positions
Word Problems
Practice
Review 2

1B

Chapter 10
Length

Comparing Lengths Directly
Comparing Lengths Indirectly
Comparing Lengths with Units
Practice

Chapter 11
Comparing

Subtraction as Comparison
Making Comparison
Subtraction Stories
Picture Graphs
Practice

Chapter 12
Numbers to 40

Numbers to 40
Tens and Ones
Counting by Tens and Ones
Comparing
Practice

Chapter 13
Addition and Subtraction Within 40

Add Ones
Subtract Ones
Make the Next Ten
Use Addition Facts
Subtract from Tens
Use Subtraction Facts
Add Three Numbers
Practice

Chapter 14
Grouping and Sharing

Adding Equal Groups
Sharing
Grouping
Practice

Chapter 15
Fractions

Halves
Fourths
Practice
Review 3

Chapter 16
Numbers to 100

Numbers to 100
Tens and Ones
Count by Ones or Tens
Compare Numbers to 100
Practice

Chapter 17
Addition and Subtraction Within 100

Add Ones — Part 1
Add Tens
Add Ones — Part 2
Add Tens and Ones — Part 1
Add Tens and Ones — Part 2
Subtract Ones — Part 1
Subtract from Tens
Subtract Ones — Part 2
Subtract Tens

Dimensions Math® Scope & Sequence

Dividing by 5 and 10
Practice C
Word Problems
Review 2

2B

Chapter 8
Mental Calculation

Adding Ones Mentally
Adding Tens Mentally
Making 100
Adding 97, 98, or 99
Practice A
Subtracting Ones Mentally
Subtracting Tens Mentally
Subtracting 97, 98, or 99
Practice B
Practice C

Chapter 9
Multiplication and Division of 3 and 4

The Multiplication Table of 3
Multiplication Facts of 3
Dividing by 3
Practice A
The Multiplication Table of 4
Multiplication Facts of 4
Dividing by 4
Practice B
Practice C

Chapter 10
Money

Making $1
Dollars and Cents
Making Change
Comparing Money
Practice A
Adding Money
Subtracting Money
Practice B

Chapter 11
Fractions

Halves and Fourths
Writing Unit Fractions
Writing Fractions
Fractions that Make 1 Whole
Comparing and Ordering Fractions
Practice
Review 3

Chapter 12
Time

Telling Time
Time Intervals
A.M. and P.M.
Practice

Chapter 13
Capacity

Comparing Capacity
Units of Capacity
Practice

Chapter 14
Graphs

Picture Graphs
Bar Graphs
Practice

Chapter 15
Shapes

Straight and Curved Sides
Polygons
Semicircles and Quarter-circles
Patterns
Solid Shapes
Practice
Review 4
Review 5

3A

Chapter 1
Numbers to 10,000

Numbers to 10,000
Place Value — Part 1
Place Value — Part 2
Comparing Numbers
The Number Line
Practice A
Number Patterns
Rounding to the Nearest Thousand
Rounding to the Nearest Hundred
Rounding to the Nearest Ten
Practice B

Dimensions Math® Scope & Sequence

The Multiplication Table of 9
Multiplying by 8 and 9
Dividing by 8 and 9
Practice B

Chapter 9
Fractions — Part 1

Fractions of a Whole
Fractions on a Number Line
Comparing Fractions with Like Denominators
Comparing Fractions with Like Numerators
Practice

Chapter 10
Fractions — Part 2

Equivalent Fractions
Finding Equivalent Fractions
Simplifying Fractions
Comparing Fractions — Part 1
Comparing Fractions — Part 2
Practice A
Adding and Subtracting Fractions — Part 1
Adding and Subtracting Fractions — Part 2
Practice B

Chapter 11
Measurement

Meters and Centimeters
Subtracting from Meters
Kilometers
Subtracting from Kilometers
Liters and Milliliters
Kilograms and Grams
Word Problems
Practice
Review 3

Chapter 12
Geometry

Circles
Angles
Right Angles
Triangles
Properties of Triangles
Properties of Quadrilaterals
Using a Compass
Practice

Chapter 13
Area and Perimeter

Area
Units of Area
Area of Rectangles
Area of Composite Figures
Practice A
Perimeter
Perimeter of Rectangles
Area and Perimeter
Practice B

Chapter 14
Time

Units of Time
Calculating Time — Part 1
Practice A
Calculating Time — Part 2
Calculating Time — Part 3
Calculating Time — Part 4
Practice B

Chapter 15
Money

Dollars and Cents
Making $10
Adding Money
Subtracting Money
Word Problems
Practice
Review 4
Review 5

4A

Chapter 1
Numbers to One Million

Numbers to 100,000
Numbers to 1,000,000
Number Patterns
Comparing and Ordering Numbers
Rounding 5-Digit Numbers
Rounding 6-Digit Numbers
Calculations and Place Value
Practice

Chapter 2
Addition and Subtraction

Addition
Subtraction
Other Ways to Add and Subtract — Part 1
Other Ways to Add and Subtract — Part 2
Word Problems

Dimensions Math® Scope & Sequence

Chapter 11
Area and Perimeter

Area of Rectangles — Part 1
Area of Rectangles — Part 2
Area of Composite Figures
Perimeter — Part 1
Perimeter — Part 2
Practice

Chapter 12
Decimals

Tenths — Part 1
Tenths — Part 2
Hundredths — Part 1
Hundredths — Part 2
Expressing Decimals as Fractions in Simplest Form
Expressing Fractions as Decimals
Practice A
Comparing and Ordering Decimals
Rounding Decimals
Practice B

Chapter 13
Addition and Subtraction of Decimals

Adding and Subtracting Tenths
Adding Tenths with Regrouping
Subtracting Tenths with Regrouping
Practice A
Adding Hundredths
Subtracting from 1 and 0.1
Subtracting Hundredths
Money, Decimals, and Fractions
Practice B
Review 3

Chapter 14
Multiplication and Division of Decimals

Multiplying Tenths and Hundredths
Multiplying Decimals by a Whole Number — Part 1
Multiplying Decimals by a Whole Number — Part 2
Practice A
Dividing Tenths and Hundredths
Dividing Decimals by a Whole Number — Part 1
Dividing Decimals by a Whole Number — Part 2
Dividing Decimals by a Whole Number — Part 3
Practice B

Chapter 15
Angles

The Size of Angles
Measuring Angles
Drawing Angles
Adding and Subtracting Angles
Reflex Angles
Practice

Chapter 16
Lines and Shapes

Perpendicular Lines
Parallel Lines
Drawing Perpendicular and Parallel Lines
Quadrilaterals
Lines of Symmetry
Symmetrical Figures and Patterns
Practice

Chapter 17
Properties of Cuboids

Cuboids
Nets of Cuboids
Faces and Edges of Cuboids
Practice
Review 4
Review 5

5A

Chapter 1
Whole Numbers

Numbers to One Billion
Multiplying by 10, 100, and 1,000
Dividing by 10, 100, and 1,000
Multiplying by Tens, Hundreds, and Thousands
Dividing by Tens, Hundreds, and Thousands
Practice

Chapter 2
Writing and Evaluating Expressions

Expressions with Parentheses
Order of Operations — Part 1
Order of Operations — Part 2

Chapter 3
Multiplication and Division

Chapter 4
Addition and Subtraction of Fractions

Chapter 5
Multiplication of Fractions

Chapter 6
Division of Fractions

Chapter 7
Measurement

Chapter 8
Volume of Solid Figures

5B

Chapter 9
Decimals

Dimensions Math® Scope & Sequence

Conversion of Measures
Mental Calculation
Practice B

Chapter 10
The Four Operations of Decimals

Adding Decimals to Thousandths
Subtracting Decimals
Multiplying by 0.1 or 0.01
Multiplying by a Decimal
Practice A
Dividing by a Whole Number — Part 1
Dividing by a Whole Number — Part 2
Dividing a Whole Number by 0.1 and 0.01
Dividing a Whole Number by a Decimal
Practice B

Chapter 11
Geometry

Measuring Angles
Angles and Lines
Classifying Triangles
The Sum of the Angles in a Triangle
The Exterior Angle of a Triangle
Classifying Quadrilaterals
Angles of Quadrilaterals — Part 1
Angles of Quadrilaterals — Part 2
Drawing Triangles and Quadrilaterals
Practice

Chapter 12
Data Analysis and Graphs

Average — Part 1
Average — Part 2
Line Plots
Coordinate Graphs
Straight Line Graphs
Practice
Review 3

Chapter 13
Ratio

Finding the Ratio
Equivalent Ratios
Finding a Quantity
Comparing Three Quantities
Word Problems
Practice

Chapter 14
Rate

Finding the Rate
Rate Problems — Part 1
Rate Problems — Part 2
Word Problems
Practice

Chapter 15
Percentage

Meaning of Percentage
Expressing Percentages as Fractions
Percentages and Decimals
Expressing Fractions as Percentages
Practice A
Percentage of a Quantity
Word Problems
Practice B
Review 4
Review 5

Suggested number of class periods: 4–5

	Lesson	Page	Resources	Objectives
	Chapter Opener	p. 3	TB: p. 1	
1	First	p. 4	TB: p. 2 WB: p. 1	Identify first from the front.
2	Second and Third	p. 5	TB: p. 3 WB: p. 3	Identify second and third.
3	Fourth and Fifth	p. 6	TB: p. 4 WB: p. 5	Identify fourth and fifth from a starting position.
4	Practice	p. 8	TB: p. 5 WB: p. 7	Practice concepts introduced in the chapter.
	Workbook Solutions	p. 11		

Chapter Vocabulary

- Order
- First
- Second
- Third
- Fourth
- Fifth
- Last
- Starting point
- Top
- Bottom
- Left
- Right

Chapters 4 through 7 of **Dimensions Math® PKA** introduced students to the cardinal numbers 1 through 10. In this chapter, students will learn about ordinal numbers.

Ordinal numbers describe position rather than quantity. Students will learn that order depends on the stated starting point. "Second from the front" and "fourth from the top" are examples.

Students should know how to count on by ones before starting this chapter. Counting back by ones will also be helpful.

There are many opportunities throughout the day to emphasize ordinal numbers. For example, "If your birthday is today, you may line up first."

Materials

- Animal counters
- Beads
- Blocks or other building materials
- Crayons
- Pipe Cleaners
- Square tiles or pieces of paper
- Stuffed toys

Note: Materials for Activities will be listed in detail in each lesson.

Storybooks

- *The Very Hungry Caterpillar* by Eric Carle
- *Henry the Fourth* by Stuart J. Murphy
- *Goldilocks and the Three Bears*

Optional Snacks

- Multi-colored cereal
- Teddy bear crackers
- Apples
- Pears
- Plums
- Green and red seedless grapes
- Celery sticks, cream cheese and fish crackers

Letters Home

- Chapter 8 Letter

Explore

Ask students where they have heard the words "first," "second," "third," "fourth," and "fifth" used. Discuss marathons and other races with them.

Learn

Have students look at the page and ask which friend is leading the line and how they know. Repeat with the airplanes.

Extend Learn

Bundle Up: Ask students to discuss their experiences in cold weather. Have they seen snow? Why are the friends wearing scarves? What other clothing could they be wearing if they are cold?

Chapter 8

Ordinal Numbers

1

Objective

- Identify first from the front.

Lesson Materials

- Blocks or other building materials, 3 different colors
- Square tiles or pieces of paper, 3 different colors
- Optional snack: multi-colored cereal (have students tell which color they ate first)

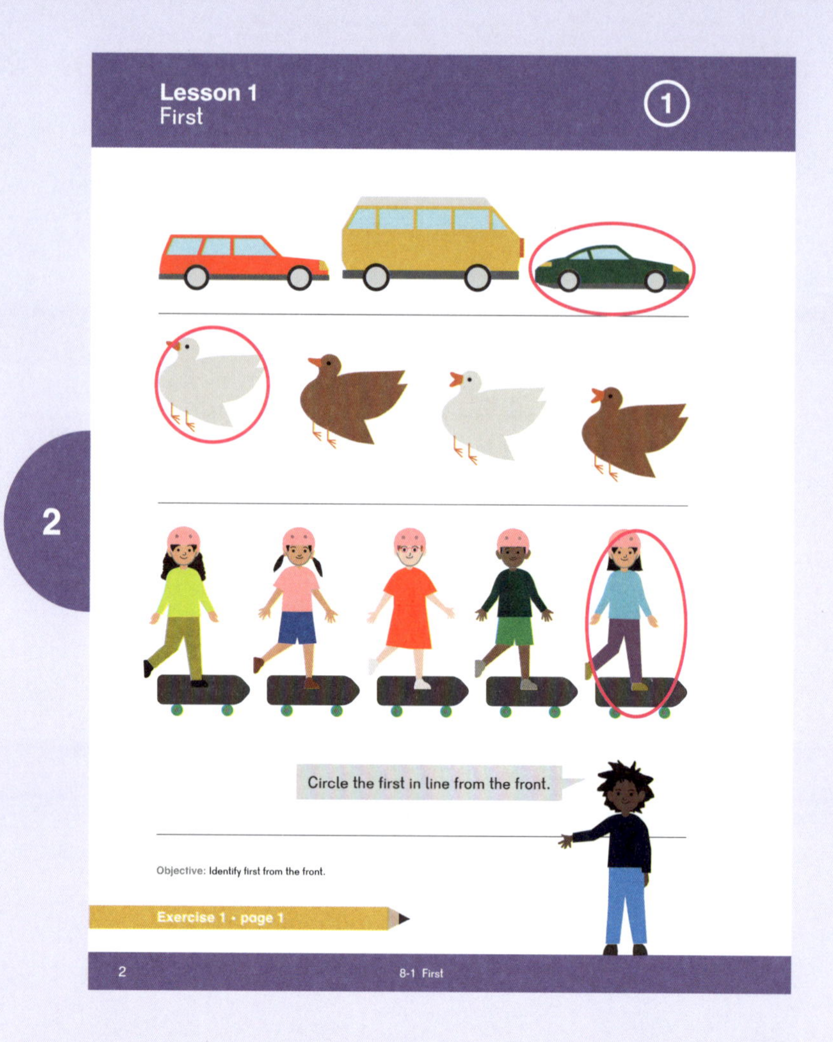

Explore

Have all students line up facing the classroom door. Ask, "Who is first in line?" Then ask, "Who is last in line?" Have students get into groups of five and stand in straight lines facing the same direction. Repeat the questions. Be sure that in instances where all objects are facing or heading in one direction, the first student is relative to the direction they are heading or facing.

Learn

Ask students which of the vehicles in the top row of page 2 is first. Have them circle the green car. Repeat with the ducks and the friends.

Whole Group Play

Have students race (run, duck walk, crab walk, etc.) and discuss which of them came in first.

Small Group Center Play

Building: Have students create a row of building materials, specifying which color block is first.

Dramatic Play: Have students, in groups of three, dress up as their character of choice, line up, and say who is first in line.

Patterns: After designating which color is first, have students create repeating patterns using two or three colors.

Exercise 1 • page 1

Extend Explore

Who is First?: With students in small groups, give clues as to which student should be first and last in their lines, such as, "Someone with long hair must be first. Someone wearing shoes that tie must be last." Have students take turns giving the clues.

Objective

- Identify second and third.

Lesson Materials

- 3 stuffed toys
- 3 animal counters per student
- Green, yellow, and orange crayons
- *The Very Hungry Caterpillar* by Eric Carle
- Optional snack: teddy graham crackers or apples, pears, and plums

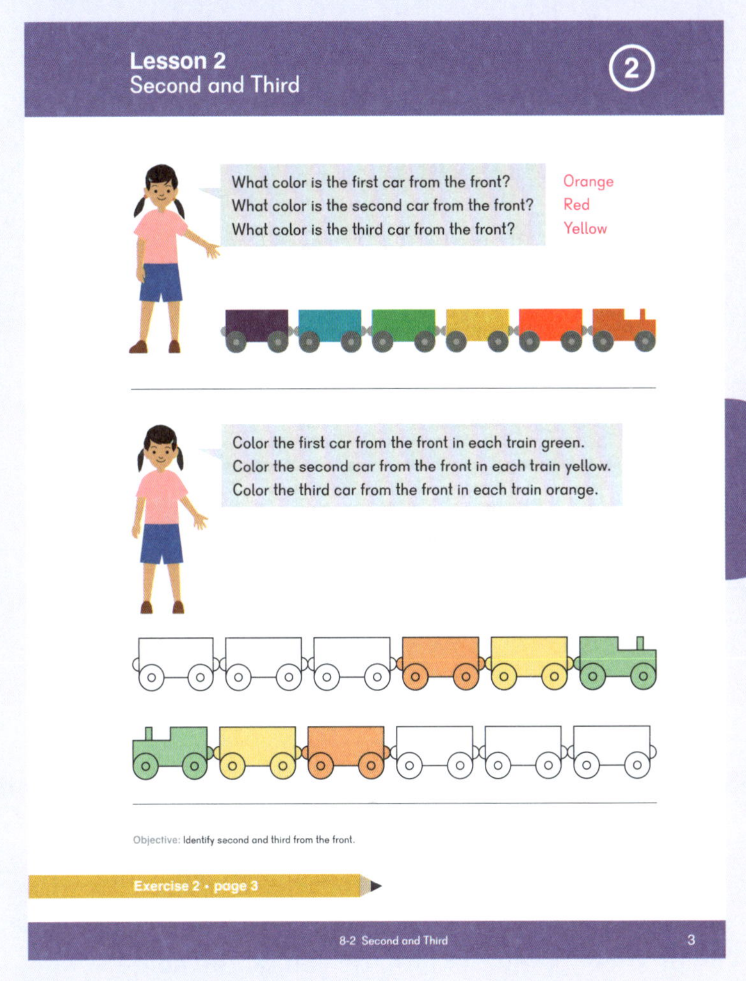

3

Explore

Tell the students a story about 3 stuffed toys going somewhere. Give them clues as to which toy was first, second, and third in line, and have them tell you how to line up the bears. Give each student 3 animal counters. Tell them another story and have them order the counters correctly.

Learn

Have students answer Sofia's questions, then color the train cars appropriately.

Whole Group Play

Have students race (hop, crawl, skip, etc.) and tell which of them came in first, second, and third.

Small Group Center Play

Dramatic Play: Have students, in groups of three, dress up as their character of choice, line up, and say who is first, second, and third in line from the front.

Patterns: After designating which colors must be second and third from the front, have students create repeating patterns using three colors.

Extend Learn

Reading Time: Read *The Very Hungry Caterpillar* by Eric Carle. Ask students to recall the first, second, and third snacks the caterpillar ate.

Point out to students that this concept of first, second, and third is different from ordinal positions. These words identify order in time.

Exercise 2 • page 3

Objective

- Identify fourth and fifth from a starting position.

Lesson Materials

- Beads, 5 different colors, 1 set of each per student
- Pipe cleaner for beads, 1 per student
- Blocks or other building materials, 5 different colors
- *Goldilocks and the Three Bears*
- Optional snack: green and red seedless grapes (have students tell which color grape they ate fifth)

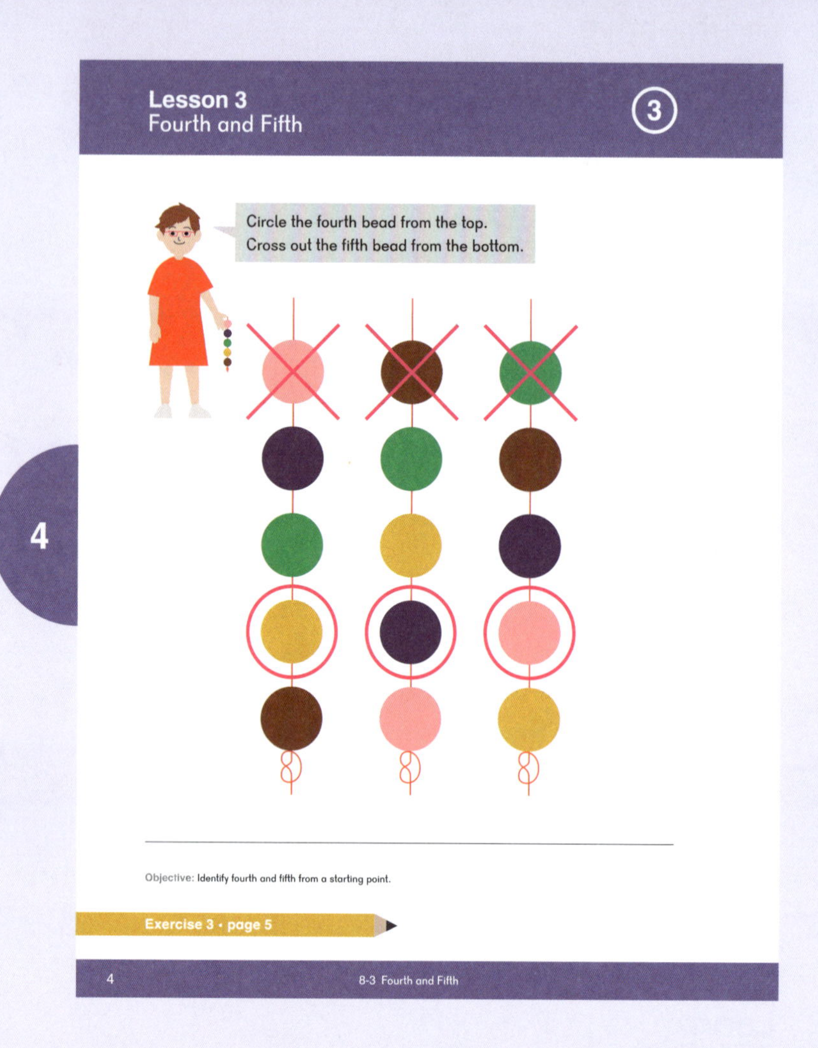

Explore

Give each student a pipe cleaner and 5 beads, each a different color. Have them put a specified colored bead on the pipe cleaner first, a different specified color second, etc., to fifth, as you do the same.

Have them hold their pipe cleaners vertically with the last bead they placed on top. Ask, "Which color bead is first from the top?" Repeat for second, third, fourth, and fifth from the top, then the bottom.

Learn

Discuss the concept of a running race, and the identification of who comes in first, second, and third. Remind them that these words (first, second, third) are used in ways other than identifying position.

Have students color and cross out beads on the page as Emma directs.

Whole Group Play

Have students race (toe-to-heel, duck walk, crab walk, etc.) in groups of 5, and tell which of them came in first through fifth.

Have students line up shoulder-to-shoulder in groups of 5. Have them name the first student from the left, second from the right, etc.

Small Group Center Play

Building: Have students create a tower using 5 different colors of building materials, then tell which one is first through fifth from the top, then from the bottom.

Patterns: Have students create repeating patterns using 3 colors, designating which colors must be second and third from the right.

Dramatic Play: Ask students which they put on first, socks or shoes. Have them give examples of other sequences in their daily lives.

Read *Goldilocks and the Three Bears* and have students act out the story. Then ask them to order the following from first to fifth:

- Goldilocks entered the house of the three bears.
- Goldilocks broke Baby Bear's chair.
- Goldilocks got lost in the woods.
- Goldilocks fell asleep in Baby Bear's bed.
- Goldilocks ate Papa Bear's porridge.

Extend Learn

Change the Story: Have students think of a different ending for *Goldilocks and the Three Bears*. Prompts include:

- What if Goldilocks was not afraid of the bears?
- What if Goldilocks decided to fix the chair instead of taking a nap?

Allow students to record their story endings.

Exercise 3 • page 5

Objective

- Practice concepts introduced in the chapter.

Lesson Materials

- *Henry the Fourth* by Stuart J. Murphy
- Red, blue, and purple crayons
- Optional snack: celery sticks, each stuffed with cream cheese and five fish crackers (Which fish is first, second, etc. from the left or right?)

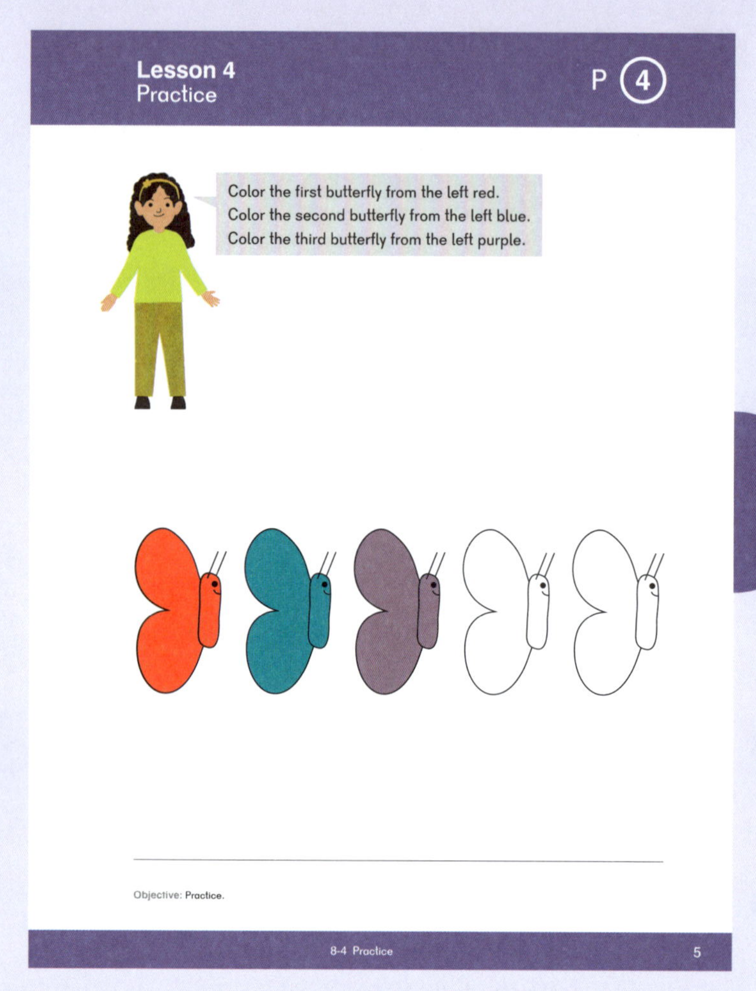

5

For the **Practice**, read the directions and speech bubbles on each page and have students complete the task.

Whole Group Play

Read *Henry the Fourth* to students.

Small Group Center Play

Building: Have students build constructions as they wish. Ask them to identify which block they placed first, second, etc.

Dramatic Play: Have students choose a character from *Henry the Fourth* and act out the story.

Dramatic Play: Lead students in a barking pattern. Tell them, for example, that the first bark will be using inside voices, the second bark will be using outside voices, the third bark will be high pitched, etc.

Exercise 4 • page 7

Extend Play

Out of Order: Have students discuss situations in which using ordinal numbers is important. Tell them that you will give them a series of directions to follow in which to complete a task. Mix up the order of the directions.

For example, give each child a piece of bread, a knife, and some butter. Tell them that there will be five directions given and that they must do things in that order.

"First, eat your bread with butter on it." They should notice that they can't do that because they have not picked up the bread, picked up the knife, dipped the knife in the butter, or spread the butter on the bread.

If time allows, have pairs of children give each other directions using ordinal numbers.

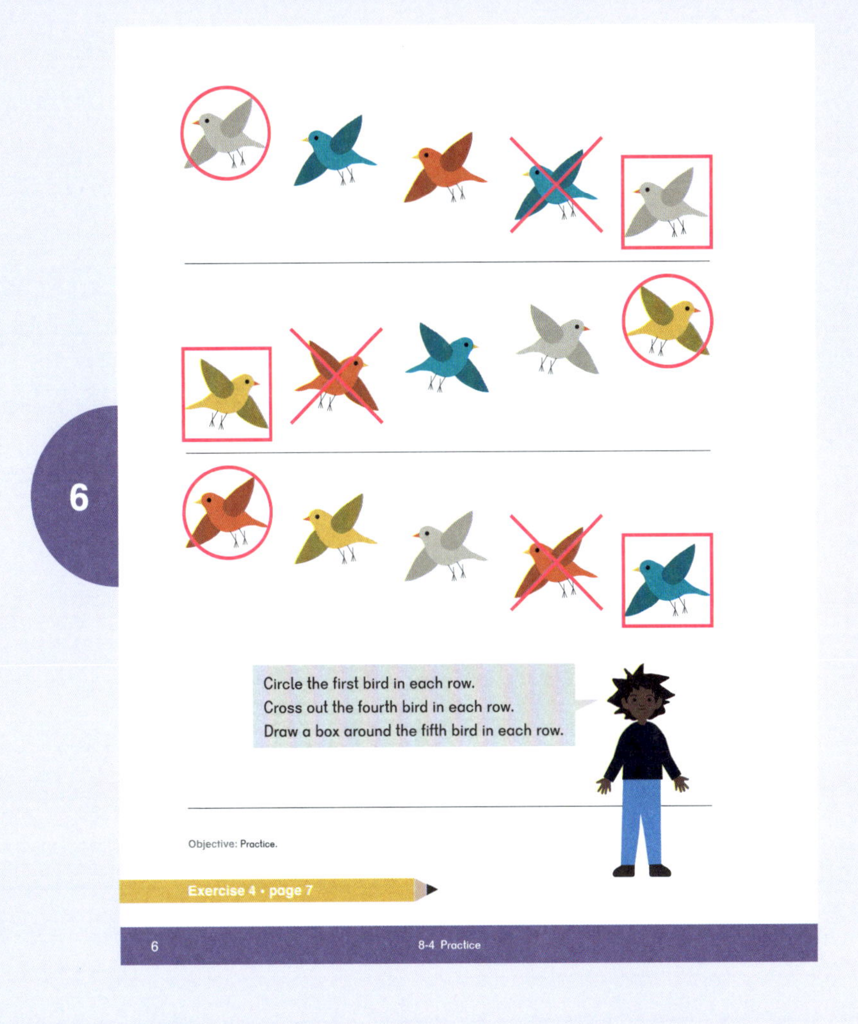

Notes

Exercise 1 • pages 1–2

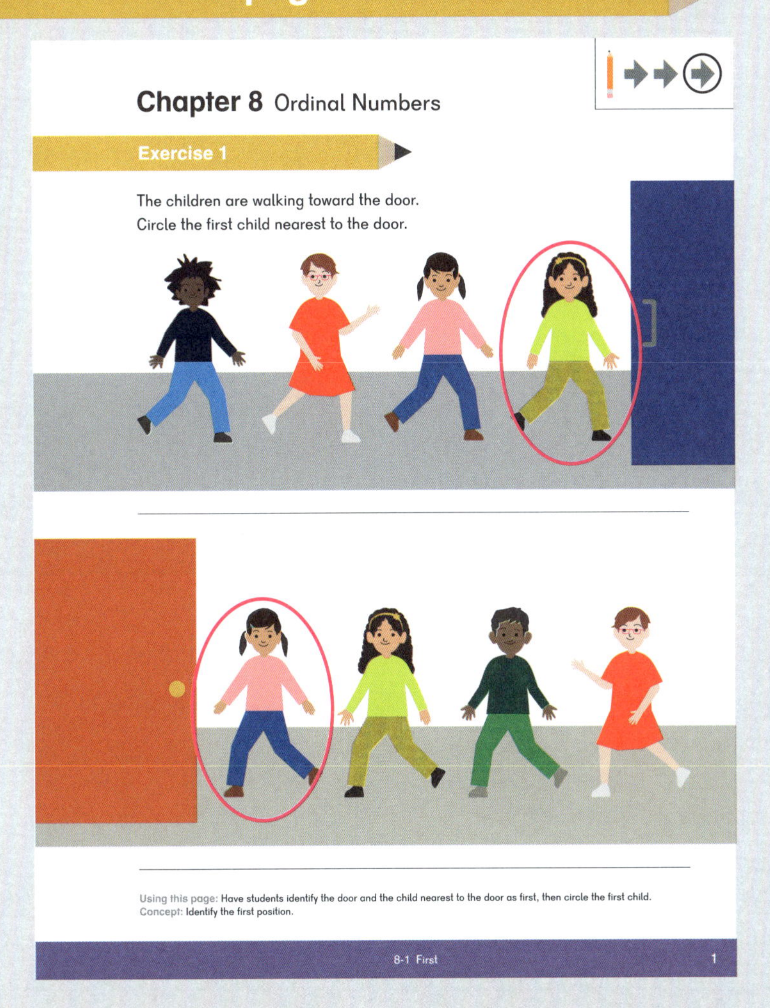

Chapter 8 Ordinal Numbers

Exercise 1

The children are walking toward the door.
Circle the first child nearest to the door.

Using this page: Have students identify the door and the child nearest to the door as first, then circle the first child.
Concept: Identify the first position.

8-1 First 1

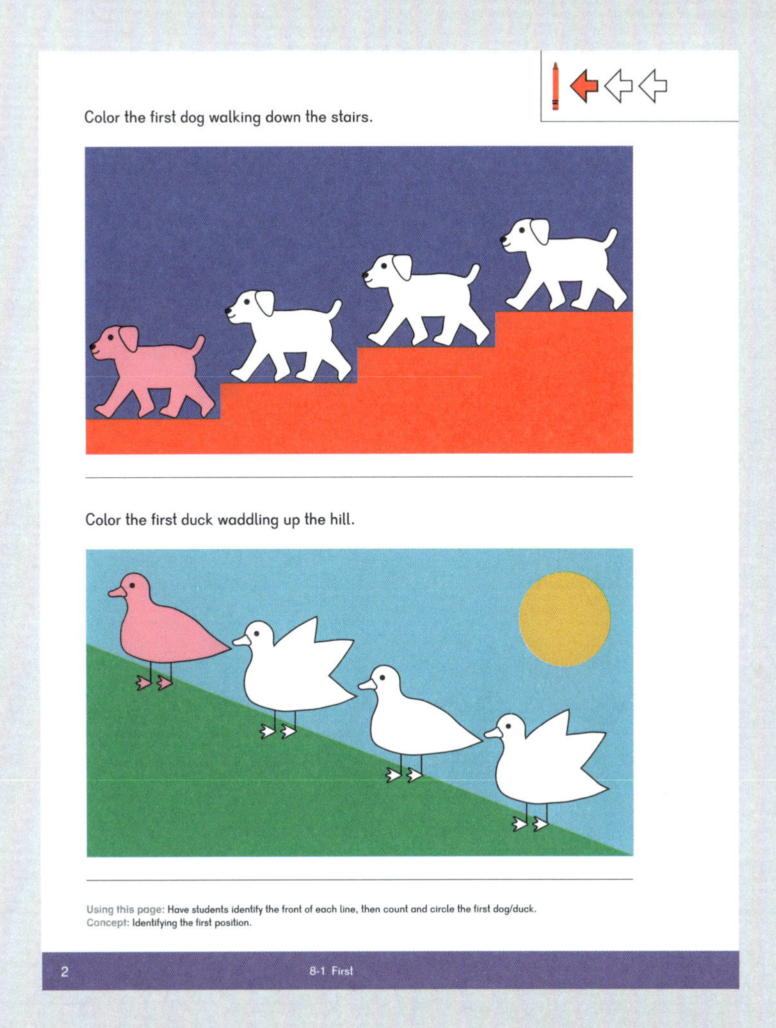

Color the first dog walking down the stairs.

Color the first duck waddling up the hill.

Using this page: Have students identify the front of each line, then count and circle the first dog/duck.
Concept: Identifying the first position.

2 8-1 First

Exercise 2 • pages 3–4

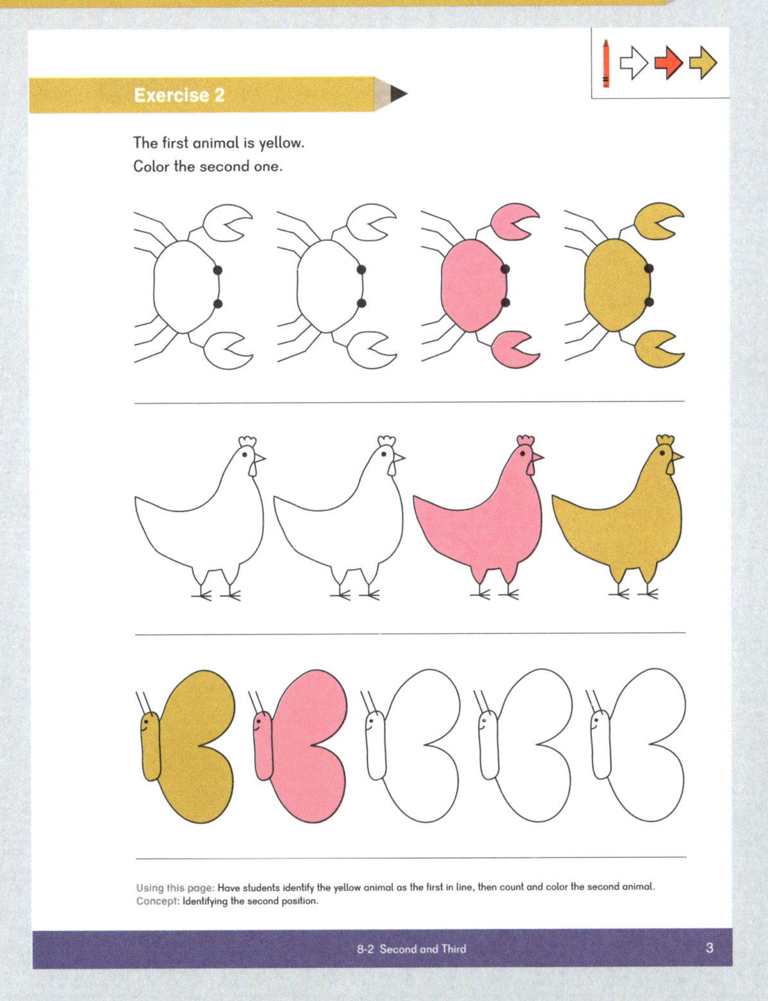

Exercise 2

The first animal is yellow.
Color the second one.

Using this page: Have students identify the yellow animal as the first in line, then count and color the second animal.
Concept: Identifying the second position.

8-2 Second and Third 3

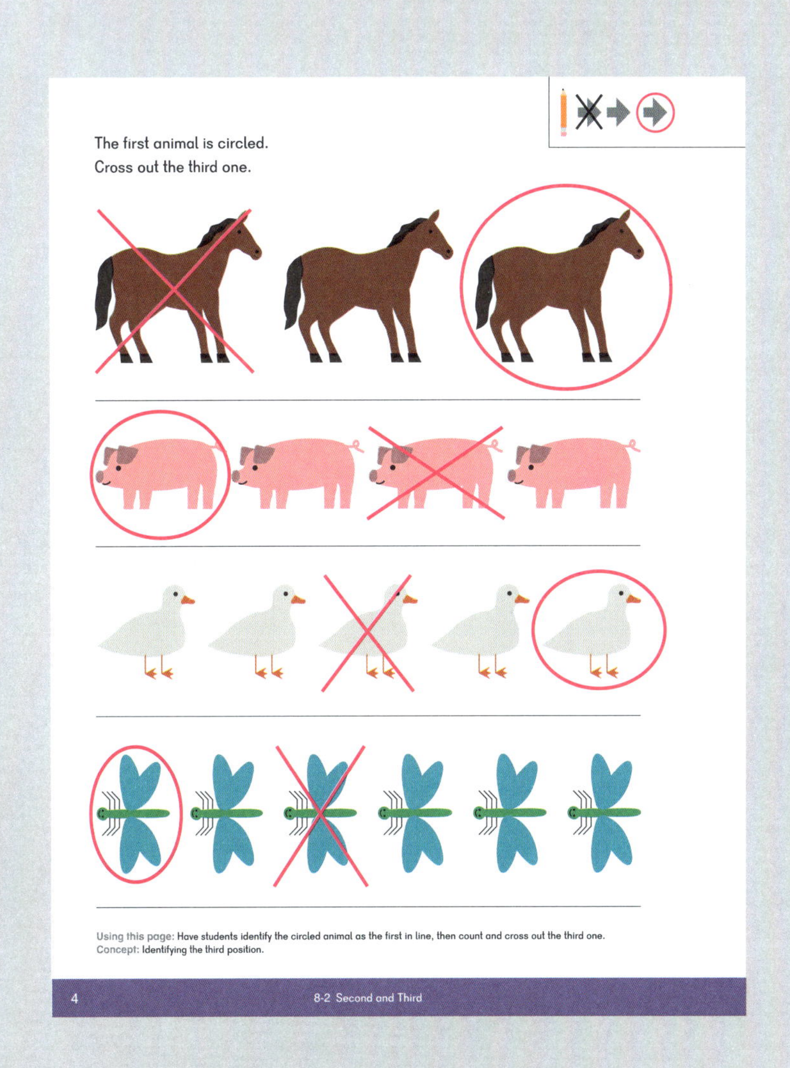

The first animal is circled.
Cross out the third one.

Using this page: Have students identify the circled animal as the first in line, then count and cross out the third one.
Concept: Identifying the third position.

4 8-2 Second and Third

Exercise 3 • pages 5–6

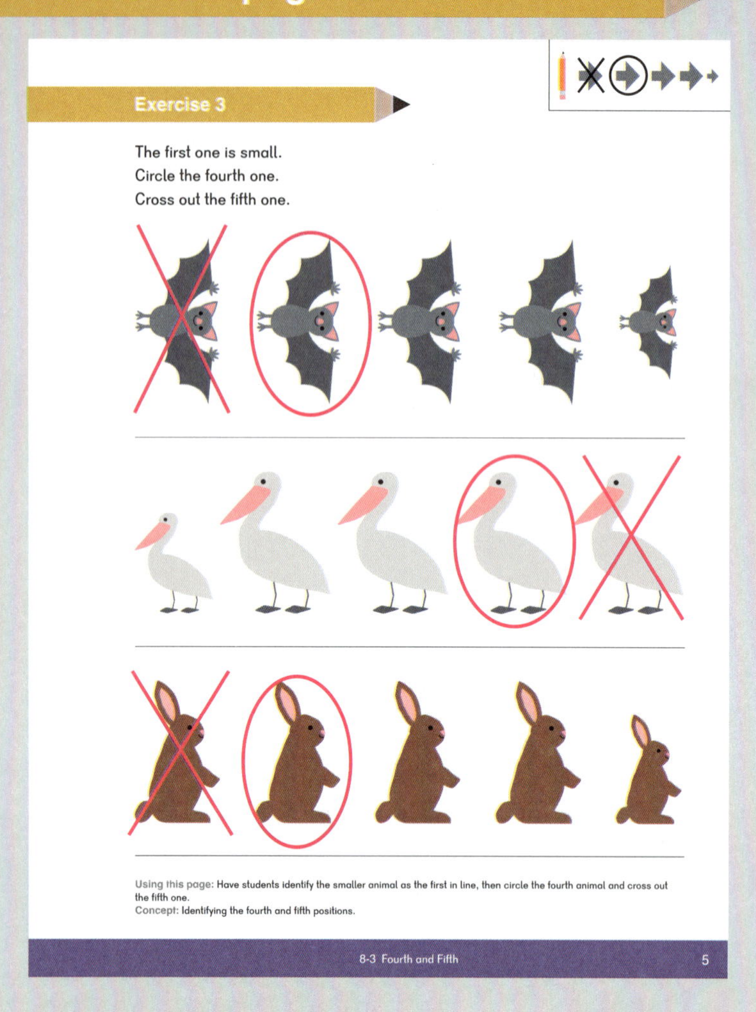

Exercise 3

The first one is small.
Circle the fourth one.
Cross out the fifth one.

Using this page: Have students identify the smaller animal as the first in line, then circle the fourth animal and cross out the fifth one.
Concept: Identifying the fourth and fifth positions.

8-3 Fourth and Fifth 5

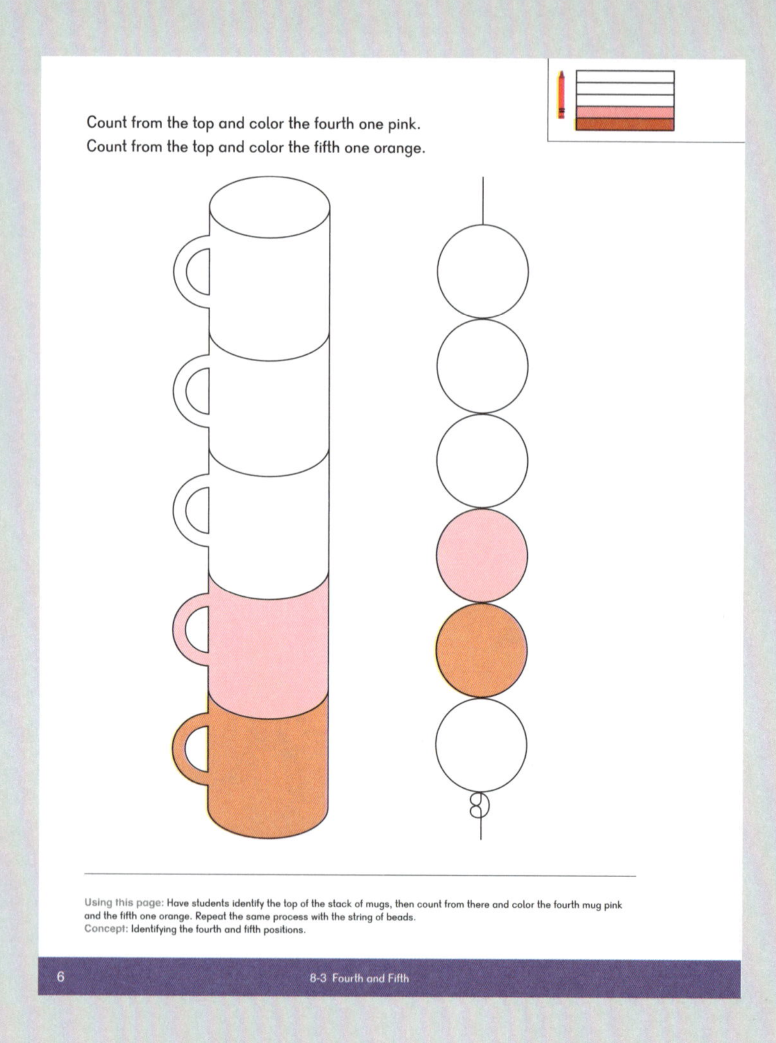

Count from the top and color the fourth one pink.
Count from the top and color the fifth one orange.

Using this page: Have students identify the top of the stack of mugs, then count from there and color the fourth mug pink and the fifth one orange. Repeat the same process with the string of beads.
Concept: Identifying the fourth and fifth positions.

6 8-3 Fourth and Fifth

Exercise 4 • pages 7–8

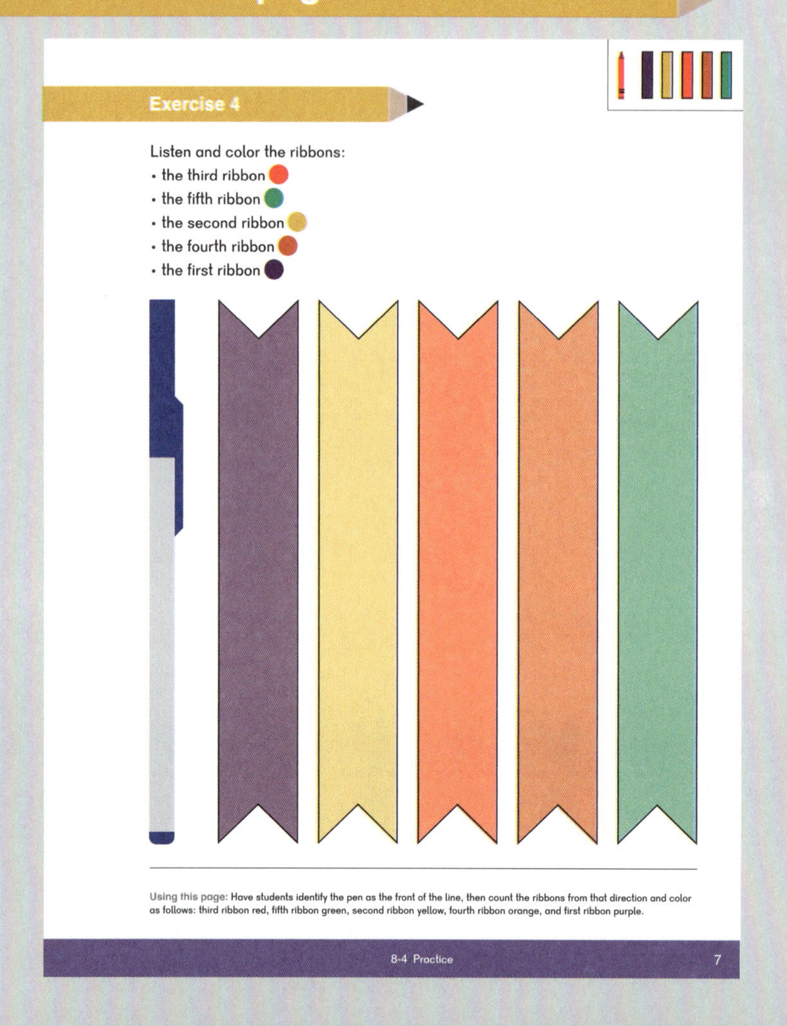

Exercise 4

Listen and color the ribbons:
- the third ribbon
- the fifth ribbon
- the second ribbon
- the fourth ribbon
- the first ribbon

Using this page: Have students identify the pen as the front of the line, then count the ribbons from that direction and color as follows: third ribbon red, fifth ribbon green, second ribbon yellow, fourth ribbon orange, and first ribbon purple.

8-4 Practice 7

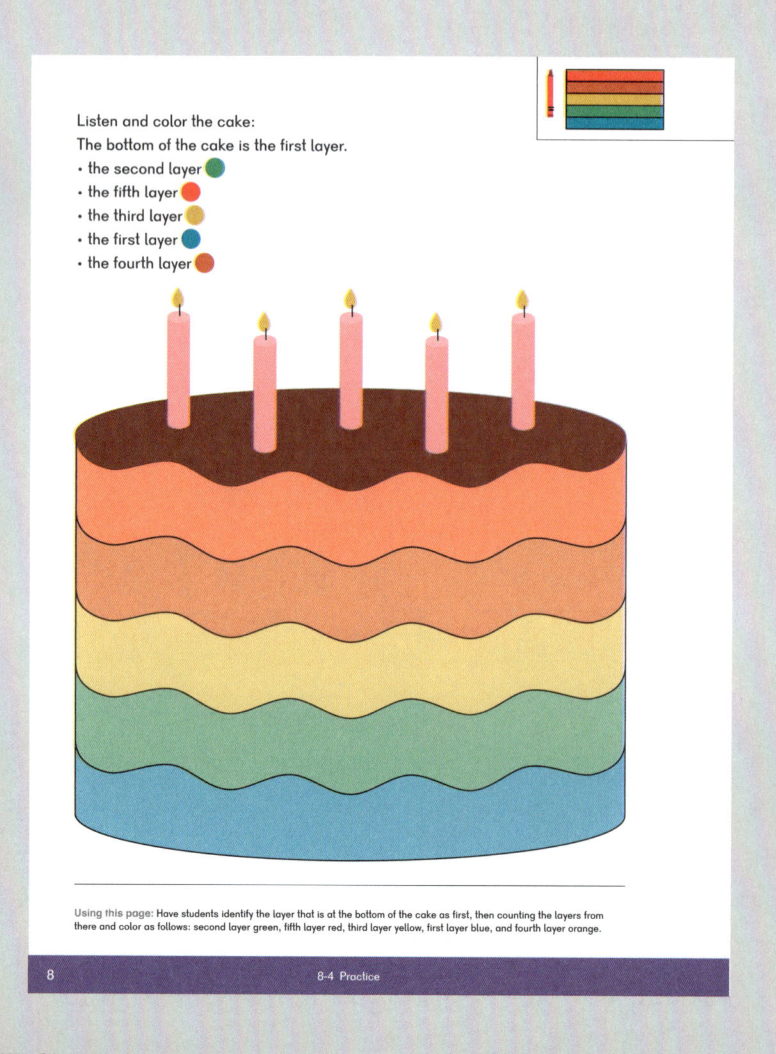

Listen and color the cake:
The bottom of the cake is the first layer.
- the second layer
- the fifth layer
- the third layer
- the first layer
- the fourth layer

Using this page: Have students identify the layer that is at the bottom of the cake as first, then counting the layers from there and color as follows: second layer green, fifth layer red, third layer yellow, first layer blue, and fourth layer orange.

8 8-4 Practice

Suggested number of class periods: 10–11

	Lesson	Page	Resources	Objectives
	Chapter Opener	p. 17	TB: p. 7	
1	Cubes, Cylinders, and Spheres	p. 18	TB: p. 8 WB: p. 9	Identify cubes, cylinders, and spheres.
2	Cubes	p. 20	TB: p. 10 WB: p. 11	Recognize attributes of cubes.
3	Positions	p. 22	TB: p. 11 WB: p. 13	Learn positional words.
4	Build with Solids	p. 24	TB: p. 13	Build using some solids.
5	Rectangles and Circles	p. 26	TB: p. 15 WB: p. 15	Recognize circles as parts of cylinders. Identify circles. Identify rectangles.
6	Squares	p. 28	TB: p. 17 WB: p. 17	Recognize squares as parts of cubes. Identify squares.
7	Triangles	p. 30	TB: p. 18 WB: p. 19	Identify triangles.
8	Squares, Circles, Rectangles, and Triangles — Part 1	p. 32	TB: p. 19 WB: p. 21	Review squares, circles, rectangles, and triangles.
9	Squares, Circles, Rectangles, and Triangles — Part 2	p. 34	TB: p. 21 WB: p. 23	Identify examples and counterexamples of circles, squares, rectangles, and triangles.
10	Practice	p. 36	TB: p. 22 WB: p. 25	Practice concepts introduced in this chapter.
	Workbook Solutions	p. 38		

Chapter Vocabulary

- Solid
- Flat
- Sphere
- Curved surface
- Flat surface
- Cylinder
- Straight
- Edge
- Shape
- Cube
- Rectangle
- Circle
- Square
- Triangle

The National Council of Teachers of Mathematics (NCTM) identified numbers, measurement, and geometry as being particularly important for students in preschool through grade 1. Research shows that children are able to form conceptual understanding of shapes in preschool.[1]

Many students enter Pre-Kindergarten able to name some of the shapes studied in this chapter. The names of the shapes will be introduced or reviewed. Attributes of the shapes and solids will be discovered and discussed. Examples and counterexamples of the shapes will be investigated, deepening students' understanding of the attributes of the various shapes.

Building with shapes will be highlighted in this chapter, allowing students to apply their understanding of mathematical concepts, plan strategies, and begin developing visualization skills.[2]

This chapter is vocabulary intensive. Continue to encourage the use of correct mathematical vocabulary. If a student is not as verbal as his or her peers, focus on conceptual understanding and assume that the vocabulary will come eventually with regular usage in class.

Ask parents to send in items shaped like cubes, cylinders, and spheres for this chapter. Toilet paper rolls, paper towel rolls, bottle caps, and oatmeal containers will be particularly helpful.

An important point covered in this chapter is that a shape does not change shape if its color, size, or orientation changes. Continue to stress this concept after it is addressed in **Lesson 5: Rectangles and Circles** and **Lesson 6: Squares**.

Paper cutouts of shapes are used in many lessons, including **Lesson 10: Practice**. Laminating the cutouts will prolong their use and save time in the future.

Positional words are in **Lesson 3: Positions**. Students learned ordinal positions in **Chapter 8: Ordinal Numbers**. In Chapter 9, students will add to that knowledge by using words such as, "in front of," "behind," "beside," "inside," "outside," etc.

Key Points

Many children believe that a square is a square, a rectangle is a rectangle, and that there is no overlap of the two shapes. The **Dimensions Math®** curriculum uses an inclusive definition of a square (as a special kind of rectangle). Students will learn this in **Dimensions Math® Kindergarten Chapter 4: Shapes and Solids**, but do not need to identify squares as rectangles at this level. It is more important that they do not identify a rectangle as a square, and that they know the length of all four sides of a square are equal.

---

[1]Gagatsis, A. & Patronis, T. (1990). Using Geometrical Models in a Process of Reflective Thinking in Learning and Teaching Mathematics. *Educational Studies in Mathematics*, Vol. 21, No. 1, 29-54.

[2]Kinzer, C., Gerhardt, K. & Coca, N. Early Childhood Educ J (2016) 44: 389. doi:10.1007/s10643-015-0717-2

Materials

- Activity cube (see page 16)
- Balloons
- Boxes or baskets
- Bubble soap (dish soap and water)
- Cardboard rolls (such as from a roll of paper towels)
- Containers to sort into
- Craft sticks
- Crayons
- Cubes of different sizes, such as dice, centimeter cubes, and blocks
- Cubes, cylinders, and spheres from geometry sets
- Flour
- Frozen paint cubes (see page 16)
- Geoboards and rubber bands
- Googly eyes
- Hula hoops
- Large chevron
- Linking cubes
- Modeling clay or play dough
- Newspaper strips
- Paint
- Paper cutouts of triangles, circles, squares, and rectangles
- Pictures of a unicycle, bicycle, and tricycle
- Pictures of art masterpieces
- Pipe cleaners
- Plastic or foam cubes of different sizes
- Poster board or large piece of art paper
- Real-life examples of cubes, cylinders, and spheres
- Ribbon
- Rounded toothpicks or coffee stirrers
- Rulers or straight edges
- Small animal counters
- Small paper cups
- Sponges or pool noodles cut into coins
- Units (cubes) from base ten block kits

Note: Materials for Activities will be listed in detail in each lesson.

Blackline Masters

- Brownie Bear Template
- Die
- Paper Cutouts — Chapter Opener
- Paper Cutouts — Lesson 5
- Paper Cutouts — Lesson 7
- Paper Cutouts — Lesson 8
- Paper Cutouts — Lesson 9

Storybooks

- *When I Build with Blocks* by Niki Alling

Optional Snacks

- Cube, cylinder, and sphere-shaped snacks such as marshmallows, grapes, or cheese cubes
- Kiwi or watermelon
- Ants on a log (celery, peanut butter, and raisins)
- Cucumber slices
- Pretzel rods
- Graham crackers
- Square crackers
- Wonton triangles
- Carrot coins
- Banana coins
- Square cheese slices

Letters Home

- Chapter 9 Letter

Bubble Wands is an activity that will also be repeated in two lessons. To create **Bubble Wands**, use pipe cleaners and drinking straws. Create circles, squares, rectangles, and triangles by bending the pipe cleaners, leaving enough length at both ends to wrap around the drinking straws.

To create an **Activity Cube**, cover each side of a large cube with a piece of paper which lists (or illustrates) an activity. Examples include: jumping jacks, hand claps, toe touches, frog leaps, high kicks, and knee slaps.

To make **Frozen Paint Cubes**, you will need an ice cube tray with cube-shaped divots, washable paints, and popsicle sticks. Start by filling about $\frac{1}{4}$ of each section of the ice cube tray with paint. Fill each section with water. Partially freeze. Add popsicle sticks, then freeze completely.

The following are lyrics to two of the songs taught in the chapter.

Position Hokey Pokey

You put one foot in.
You take one foot out.
You put one foot in and you shake it all about.
You do the hokey pokey and you turn yourself around.
That's what it's all about.

You put one hand in.
You take one hand out.
You put one hand in and you shake it all about.
You do the hokey pokey and you turn yourself around.
That's what it's all about.

You put one elbow in.
You take one elbow out.
You put one elbow in and you shake it all about.
You do the hokey pokey and you turn yourself around.
That's what it's all about.

You put your whole self in.
You take your whole self out.
You put your whole self in and you shake it all about.
You do the hokey pokey and you turn yourself around.
That's what it's all about.

Draw a Square

Start at a pointy spot
Use your pencil, make a dot
Draw straight down then to the side
Turn the corner
Up you glide
One more turn to meet the dot
All sides same length—a square you've got.

Lesson Materials

- Cubes, cylinders, and spheres from geometry sets — 1 of each solid per pair of students
- Paper Cutouts — Chapter Opener (BLM)
- Objects in similar shapes as the solids listed above, such as cubic boxes of different sizes, dice, soup cans, and different sizes of balls
- Containers for sorting, 3 per center

Explore

The objective of this **Chapter Opener** is to identify "flat" versus "solid" objects. The **Dimensions Math®** series refers to flat objects as shapes, and three-dimensional objects as solids.

Give each pair of students a cube, cylinder, sphere, and Paper Cutouts — Chapter Opener (BLM). Allow students to explore the materials for a few minutes.

Introduce the words "flat" and "solid" by holding up a cube and a square. Have students do the same. Hold up different examples of solids and paper cutouts and have students tell if the object is a solid or if it is flat.

Hold up a geometric representation of each of the three solids one at a time, name it, and discuss its attributes. Does it have any flat surfaces? Remind them that a square is flat. If so, how many? Are there edges, points, or curved surfaces? Then have students compare the different shapes. For example, both a cylinder and a sphere have curved surfaces, but a cube does not.

Learn

Have students look at page 7 and discuss the picture. Ask them to name any shapes or solids that they recognize.

Extend Explore

Mystery Containers: Create mystery containers, each holding a sphere, cylinder, and cube, one container per small group. Cover each container with a piece of cloth.

Call out the name of one of the solids. In each group, one student reaches into that group's container and pulls out that solid without looking. Repeat until all students have had a turn.

Objective

- Identify cubes, cylinders, and spheres.

Lesson Materials

- Cubes, cylinders, and spheres from geometry sets — 1 of each solid per pair of students
- Objects shaped similar to the solids listed above, such as cubic boxes of different sizes, dice, soup cans, and different sizes of balls
- Containers to sort into
- Cubes of different sizes, such as dice, centimeter cubes, and blocks
- Cardboard rolls (such as from a roll of paper towels)
- Optional snack: snacks in the shapes of cubes, cylinders, and spheres, such as cheese cubes, gelatin cubes, marshmallows, turkey and cheese pinwheels, grapes, and puff cereal

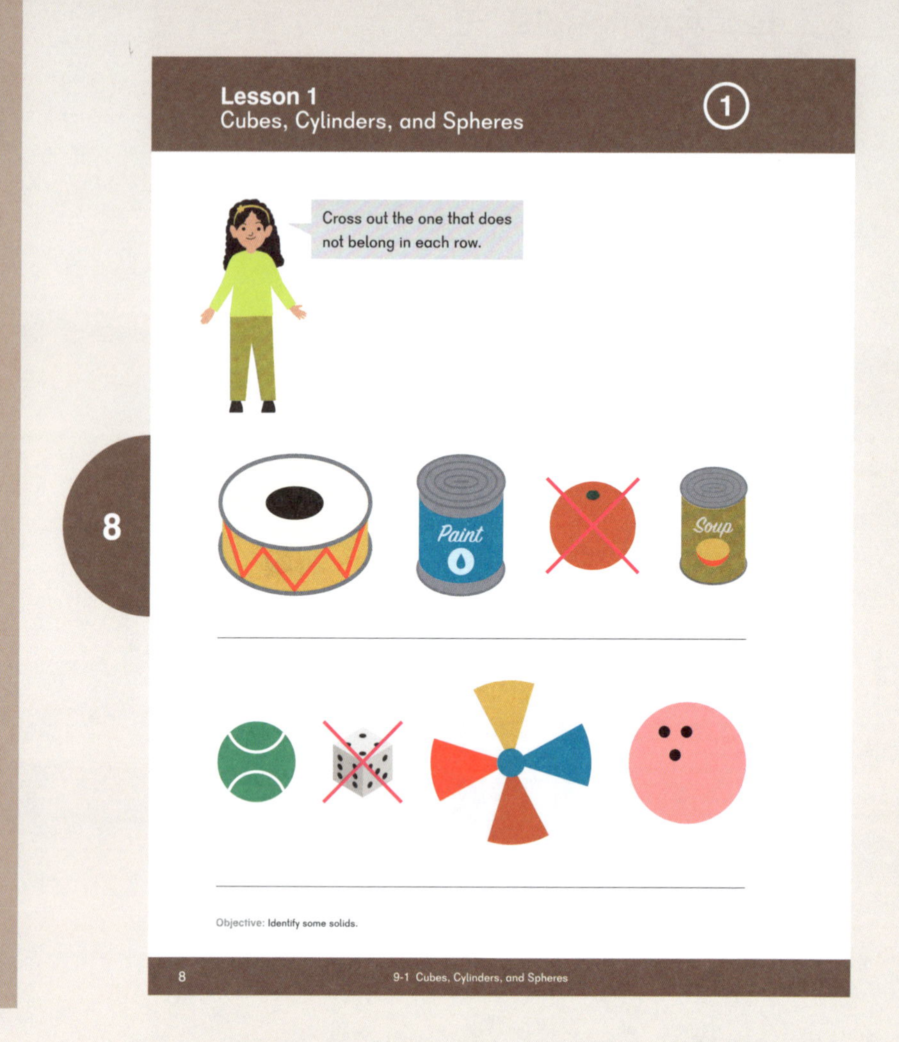

Explore

Provide examples of the solids to each pair of students. Have them determine which of the solids roll. Discuss why some of the solids roll and others do not. Remind them of the term "curved surface."

Have students help you sort cubes, cylinders, and spheres into three groups. Start by putting one of each solid in front of you, with space between. Hold up another cube and say, "I want to put the same types of solids together. Where should this one go?" Repeat with other solids.

Hold up examples of spheres, cylinders, and cubes in random order and have students name the solid. Then call out the name of a solid and have students hold up an example.

Set out a row of four real-life examples of solids, three of which are similar and one of which does not belong with the other three. For example, a cereal box, die, gift box, and a soccer ball. Ask students to tell which does not belong. Repeat with other objects.

Ask students what is alike and what is different about spheres and cylinders. Alike: both have curved surfaces. Different: A sphere has no flat surfaces. A cylinder has two flat surfaces.

Learn

Have students look at textbook page 8. Ask them to identify the objects in the first row (drum, paint can, orange, soup can). Have them tell which of the objects does not belong in the row, and why. (The orange is not a cylinder so it does not belong.) Repeat for the other row.

Have students look at page 9. Read Sofia's speech bubble to them and have them complete the task.

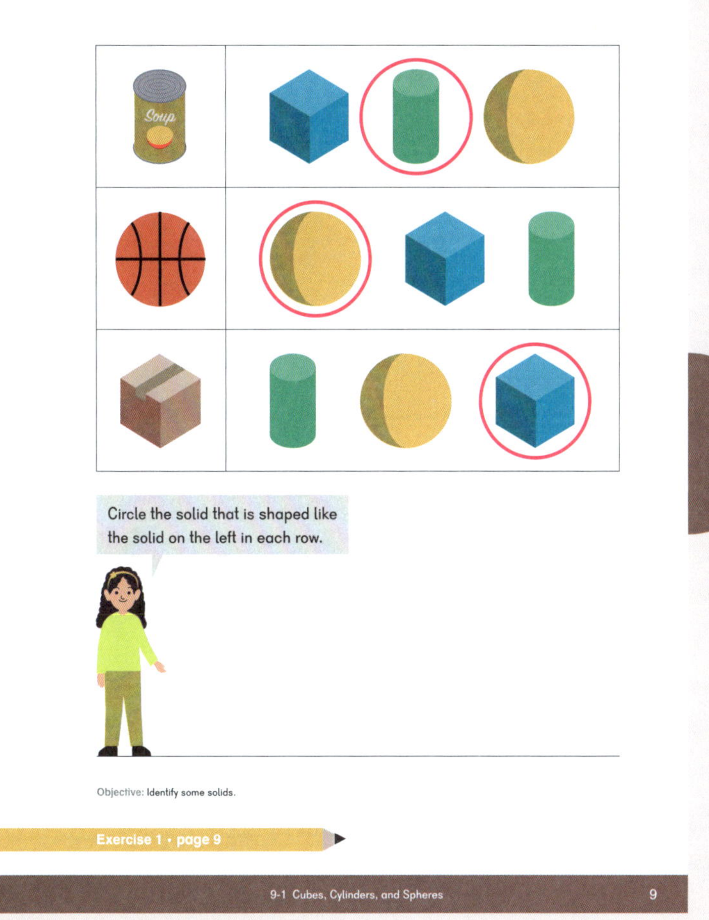

Whole Group Play

Pass the Solid: Have students sit in a circle. Show them an example of a cube, sphere, and cylinder from a geometry set. Have one student select a solid and keep it hidden from other students. Start some music, and have that student walk around the outside of the circle until the music stops. At that time, she will place the solid behind the student nearest to her and tap that student on the shoulder.

Play music again for the seated student to reach behind them and feel the solid without looking. When the music stops, the student who was tapped on the shoulder must identify it. For example, he might feel the solid and say, "It's a sphere because it has no flat surfaces." Repeat until all students have had a chance to identify a solid.

Small Group Center Play

Sort: Set up a center with numerous examples of spheres, cylinders, and cubes, and three sorting containers. Put an example of each solid in one of the containers. The task is to sort the solids into the proper containers.

Building: Provide many different sizes of cubes, such as dice, centimeter cubes, and building blocks. The task is to build the tallest tower possible with the cubes.

Cylinder Art: Set up a center with cardboard rolls, glue, feathers, glitter, and crayons. Have students create cylinder masterpieces.

Exercise 1 • page 9

Extend Play

Have students discuss why, when building a tower of cubes of different sizes, starting with particular cubes will result in a higher tower that doesn't topple than starting with other cubes. (Start with large cubes and end with small ones. Some students may want to discuss the force of gravity and how that affects construction.)

Objective

- Recognize attributes of cubes.

Lesson Materials

- Cubes from geometry sets
- Units (cubes) from base ten block kits
- Real-life examples of cubes of different sizes and colors
- Frozen paint cubes (see page 16)
- Activity cube (see page 16)
- Die (BLM) (pre-cut)
- Optional snack: kiwi or watermelon, cut in the shape of cubes

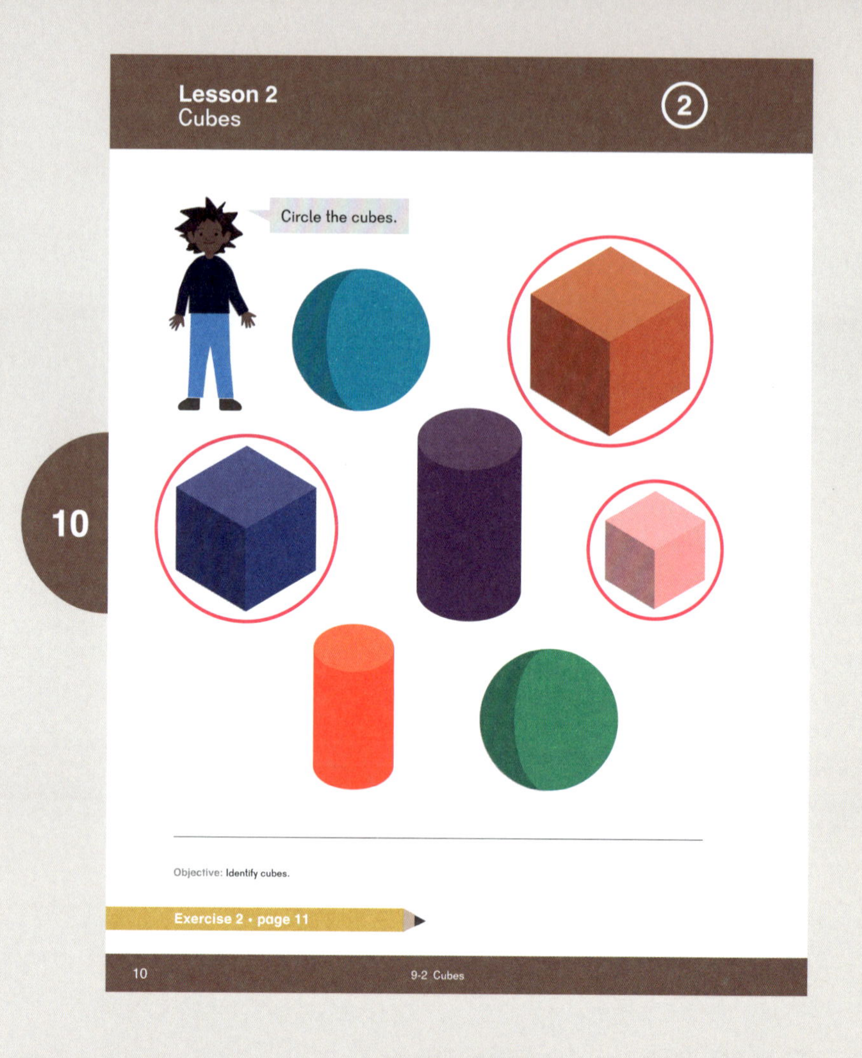

Explore

Hold up a cube and have students name it. Have them tell what they know about cubes. Discuss the flat surfaces and corners on a cube.

Set out many examples of cubes of different sizes, colors, and materials (plastic, wood, cardboard, etc.). Ask students if each is a cube. Be sure they understand that size, color, etc. are not attributes to consider in order to answer the question.

Sort the cubes and have students explain the reason for your sort. Then have students suggest different sort reasons and help you sort the cubes in different ways.

Learn

Have students look at page 10 and name all of the solids. Then have them circle the cubes.

Whole Group Play

Have students stand in a circle. Roll the **Activity Cube** and do the activity. Continue to roll it until all of the activities have been done.

Small Group Center Play

Building: Give each student a Die (BLM). Model for them how to fold on the lines and tape the edges to create a cube.

Frozen Painting: Give each student a frozen paint cube and a piece of art paper. Tell them to paint quickly and trade colors with a friend.

Exercise 2 • page 11

Extend Play

Activity Cube: Repeat the **Activity Cube** game, this time designating how many of each activity will be done each time.

Counting with a Cube: Have students identify the number of corners and faces on the cubes they built. Listen to their strategies of how to count each corner and face without counting it twice.

Objective

- Learn positional words.

Lesson Materials

- Small animal counters
- Small paper cups
- Linking cubes of 5 different colors in a tower, 1 tower per student
- Brownie Bear Templates (BLM)
- Optional snack: ants on a log

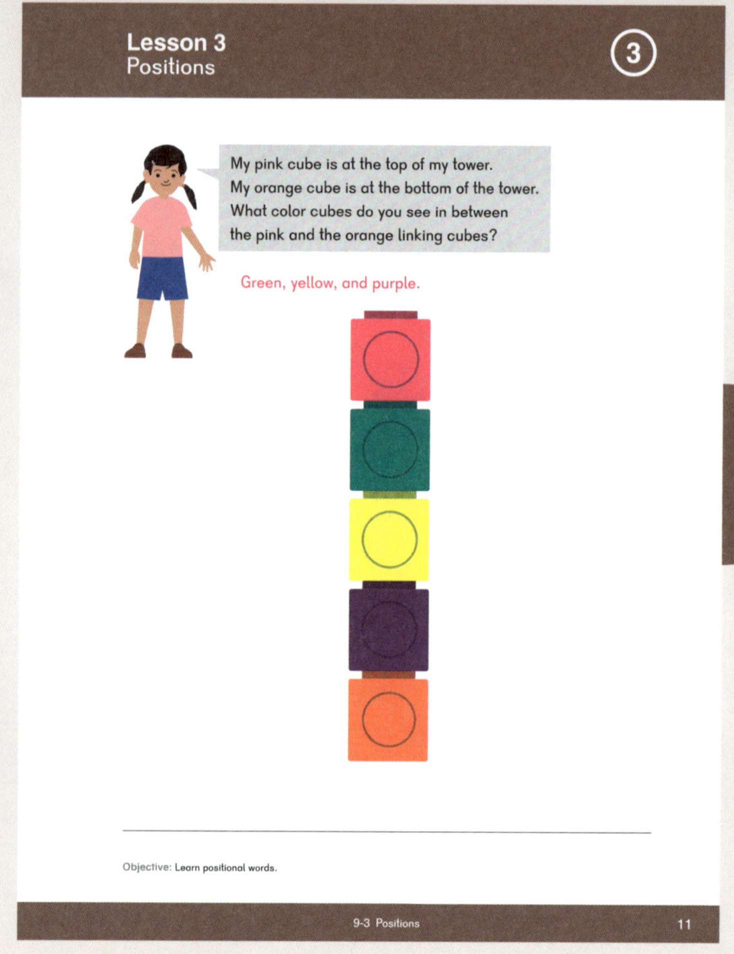

11

Explore

Give each child an animal counter and a small paper cup. Give them two minutes to play with their counters with a partner, having the animals change location and describing where their animals moved.

Tell them that they will be training their animals using their paper cups. Give them directions for where the animals must be placed using positional words. Be sure to include these terms:

- In front of
- Behind
- Beside
- Inside
- Outside
- On
- Under
- Between
- Next to
- Above
- Below

Other positional words may be included.

Collect the animal counters and cups.

Learn

Have students look at page 11 and describe the tower on the page using positional words. Then have them answer Mei's question. Build on prior knowledge from Chapter 8 by asking students to identify which cube is second from the bottom, etc.

Have students look at page 12. Read Emma's directions and have students color the blocks the appropriate colors. Ask them to say another word that Emma could have used other than "blocks" (cubes).

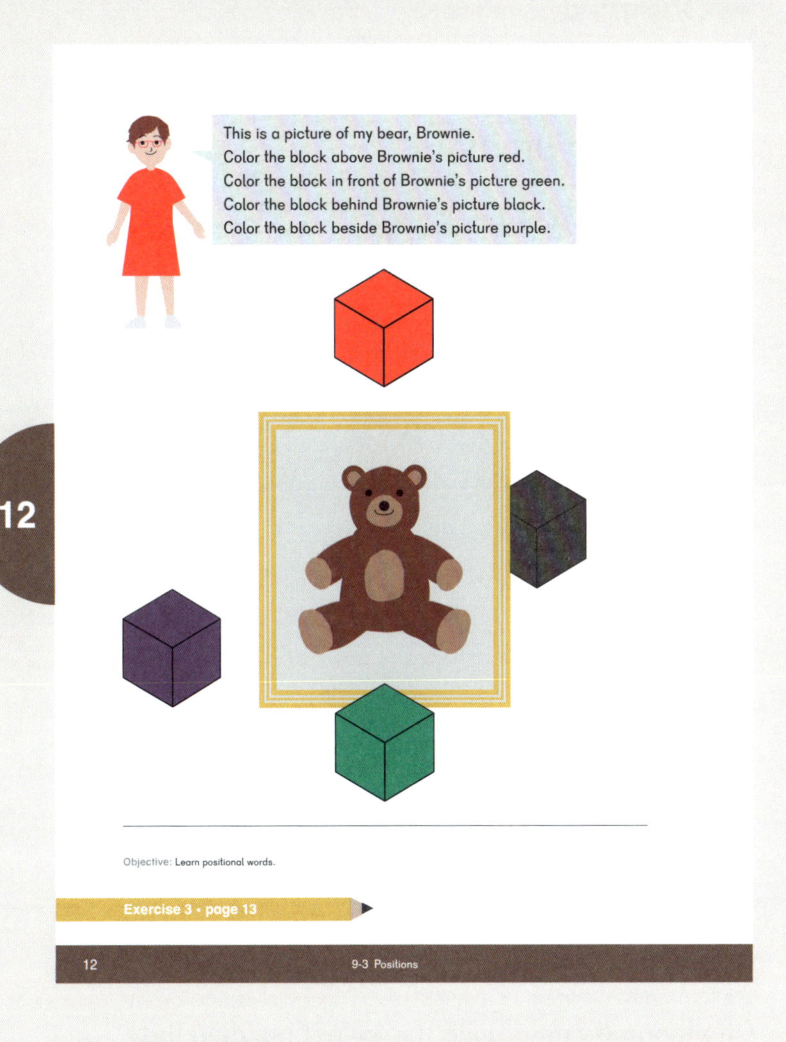

Whole Group Play

Position Hokey Pokey: Make sure that students know the meanings of "hand," "foot," and "elbow." Have students stand in a circle and do the Position Hokey Pokey (lyrics on page 16).

Small Group Center Play

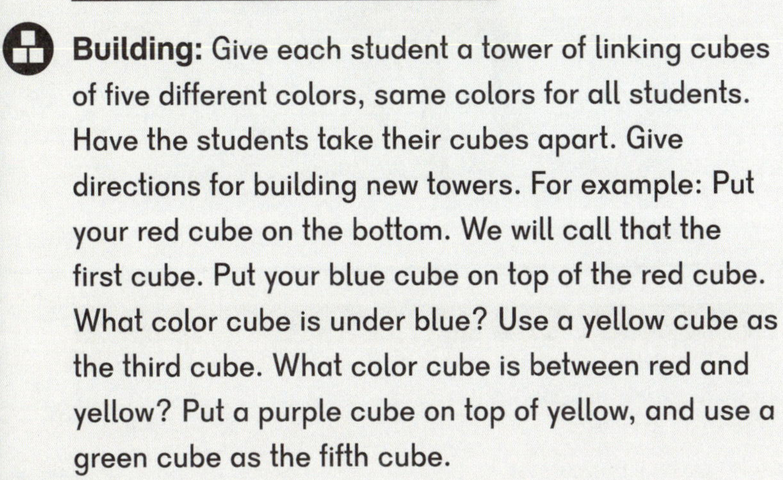

Building: Give each student a tower of linking cubes of five different colors, same colors for all students. Have the students take their cubes apart. Give directions for building new towers. For example: Put your red cube on the bottom. We will call that the first cube. Put your blue cube on top of the red cube. What color cube is under blue? Use a yellow cube as the third cube. What color cube is between red and yellow? Put a purple cube on top of yellow, and use a green cube as the fifth cube.

Art: Provide a Brownie Bear Template (BLM) and crayons to each student. Give directions for coloring the page. Examples: Draw a red line over Brownie. Draw a purple squiggle under one of Brownie's feet.

Exercise 3 • page 13

Extend Play

Play **Position Hokey Pokey** adding other parts of the body: shoulder, ankle, wrist, knee, etc.

Objective

- Build using some solids.

Lesson Materials

- Different examples of cubes, cylinders, and spheres, at least 4 of each per pair of students
- Round balloons, already blown up and tied
- Flour, water, and newspaper strips
- *When I Build with Blocks* by Niki Alling
- Optional snack: thick cucumber slices and pretzel rods

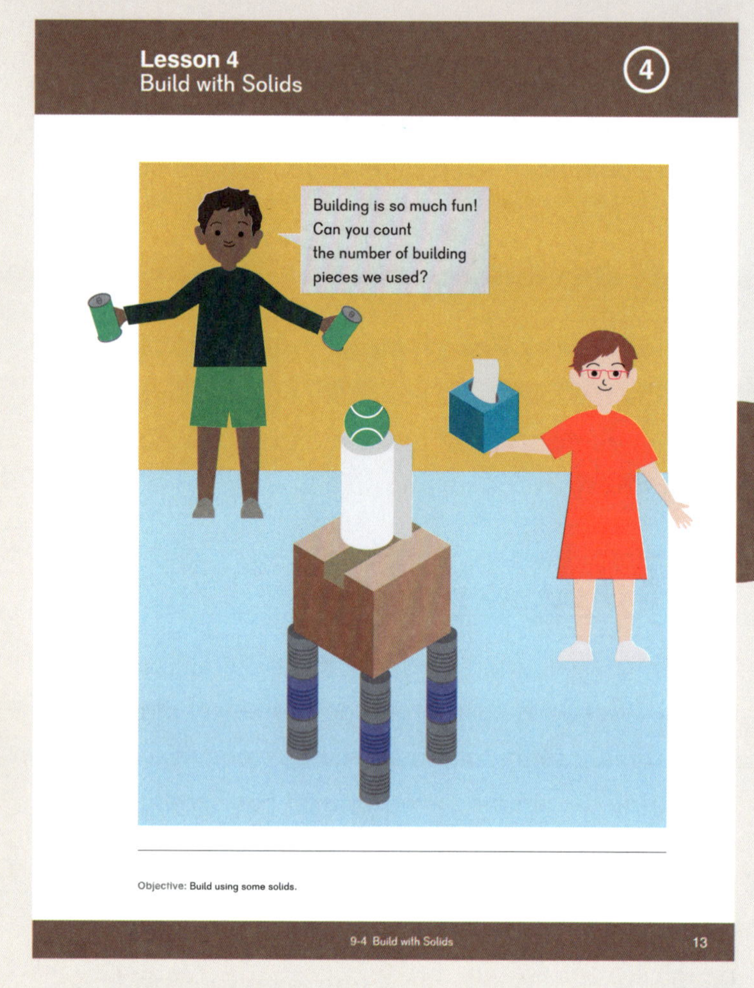

Explore

Provide at least 4 different examples of cubes, cylinders, and spheres to each pair of students.

Give students a few minutes to explore their materials as desired. Then have them name and describe the solids.

Challenge them to build the tallest building they can with their solids. Visit each pair and have them describe their buildings using positional words. Then have them build the tallest building possible using exactly 10 solids. Have them describe their buildings using positional words to other students.

Learn

Have the students look at page 13 and name the solids on the page.

Have the students look at page 14 and follow Mei's directions.

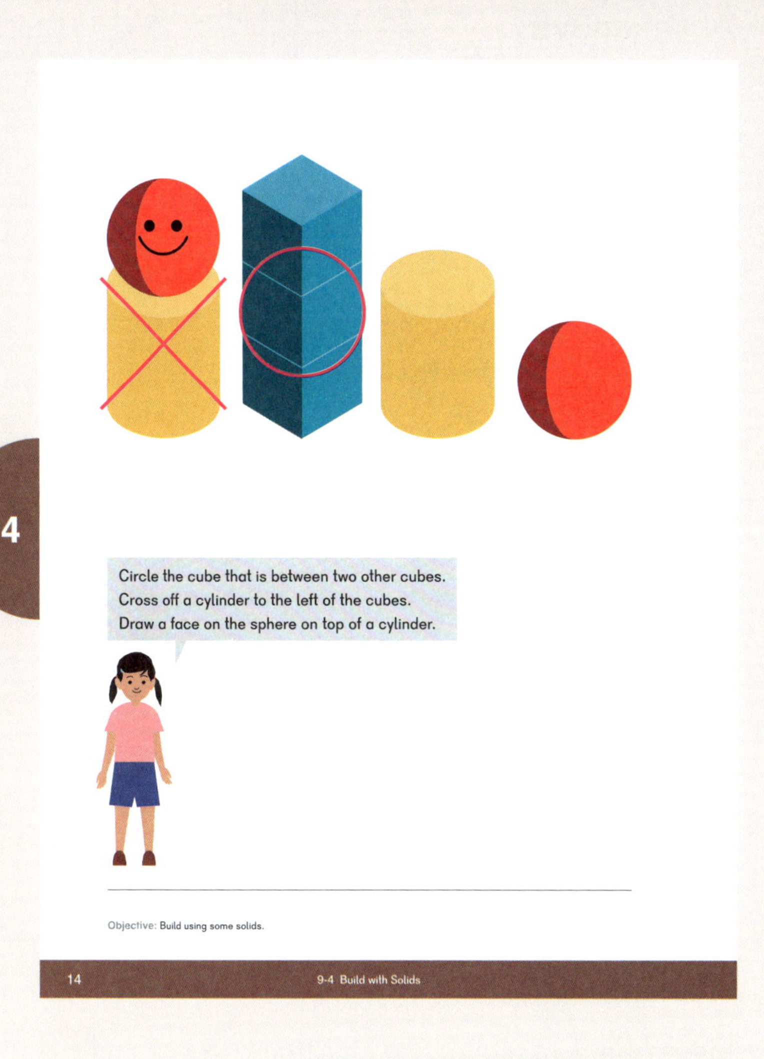

Whole Group Play

Tower of Shoes: Have students remove their shoes and build a structure as tall as possible using their shoes. After the structure is built, encourage students to identify the location of their shoes using positional words. For example, "My shoe is on top of Alyssa's shoe and next to Diego's shoe."

Small Group Center Play

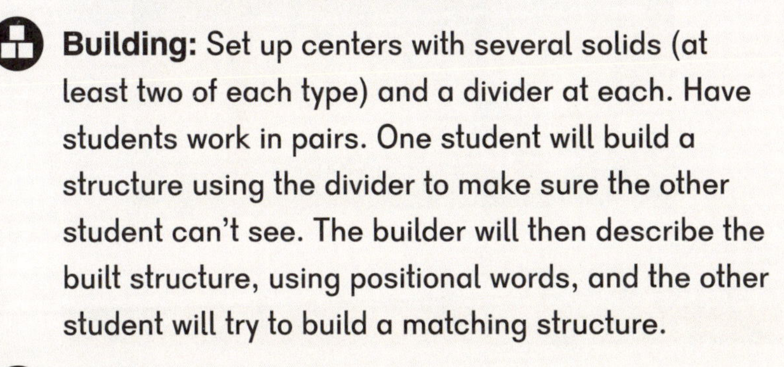

Building: Set up centers with several solids (at least two of each type) and a divider at each. Have students work in pairs. One student will build a structure using the divider to make sure the other student can't see. The builder will then describe the built structure, using positional words, and the other student will try to build a matching structure.

Papier-Mâché Spheres: Mix flour and water in a 1 : 1 ratio to create papier-mâché glue. Give each child a balloon. Have them dip newspaper strips into the glue, wipe off the excess, and wrap the strips around the balloon. They should make three or four layers of newspaper. Let the glue dry overnight. Before the next class starts, pop the balloons. They will decorate their spheres during the next class.

No exercise for this lesson

Extend Explore

Reading Time: Read *When I Build with Blocks* by Niki Alling to the students. Build a complicated structure using building blocks and have students replicate your structure.

Objective

- Recognize circles as parts of cylinders.
- Identify circles.
- Identify rectangles.

Lesson Materials

- Cylinders from geometry sets
- Real-life examples of cylinders
- Round sponges or pool noodles cut into pieces
- Paint
- Rectangular poster board or large piece of art paper
- Purple and black crayons
- Paper Cutouts — Lesson 5 (BLM)
- Rulers or straight edges, 1 per student
- Hula hoops, 1 per pair of students
- Optional snack: graham crackers

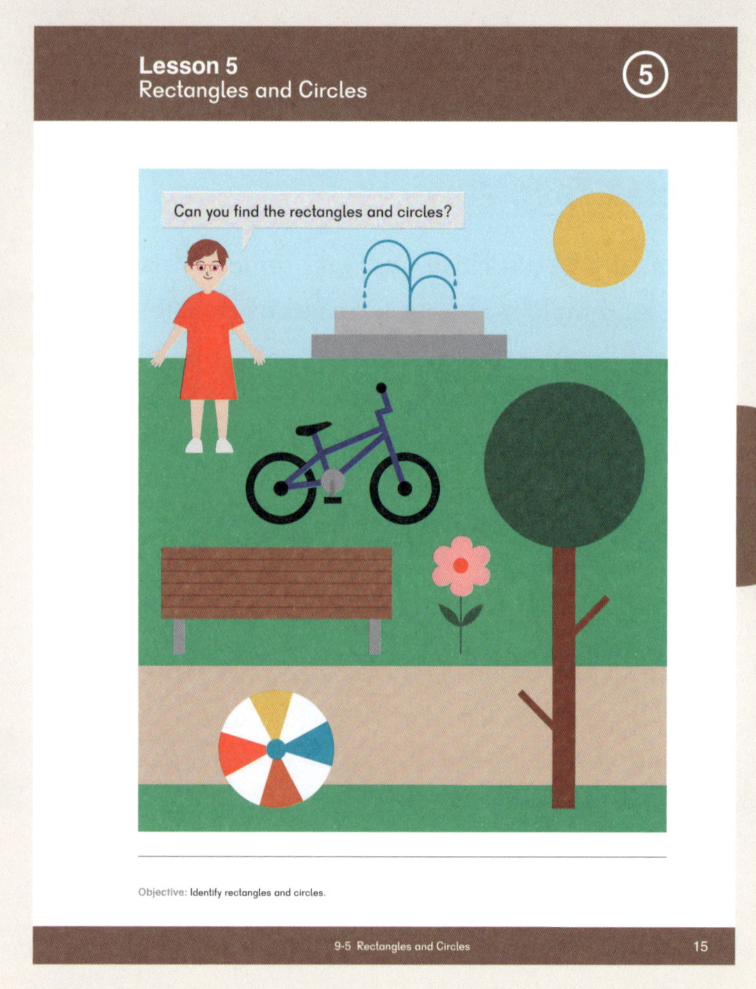

15

Explore

Place cylinders around the classroom. Send students on a treasure hunt to find cylinders and bring them back to the group.

Once all students are back in the group, show them a round sponge or pool noodle section and have them identify it as a cylinder. Then dip the sponge or noodle section into paint and transfer the paint to the poster board or art paper. Ask students to name the shape you painted. If necessary, introduce the word "circle." Have them describe a circle by answering such questions as:

- Is this shape flat or solid?
- Do you see any corners on this shape? How many?
- Are the sides of the shape straight or curved? (Trace the outside of the circle with your finger as you ask this question.)

Next, ask students to name the shape of the poster board. If necessary, introduce the word "rectangle." Show examples of different sizes of rectangles and have students describe a rectangle by answering questions such as those above. Be sure they understand that a rectangle has four sides, and that the sides are straight. Then have them discuss the differences between a rectangle and a circle.

Hold up a long, thin rectangle so that it looks tall. Ask students to name the shape. Turn the rectangle 90 degrees and ask again. Show a shorter, thicker rectangle and have them name it.

Learn

Have students look at page 15 and find the circles and rectangles.

Have students look at page 16. Read Sofia's directions and have them color the shapes appropriately.

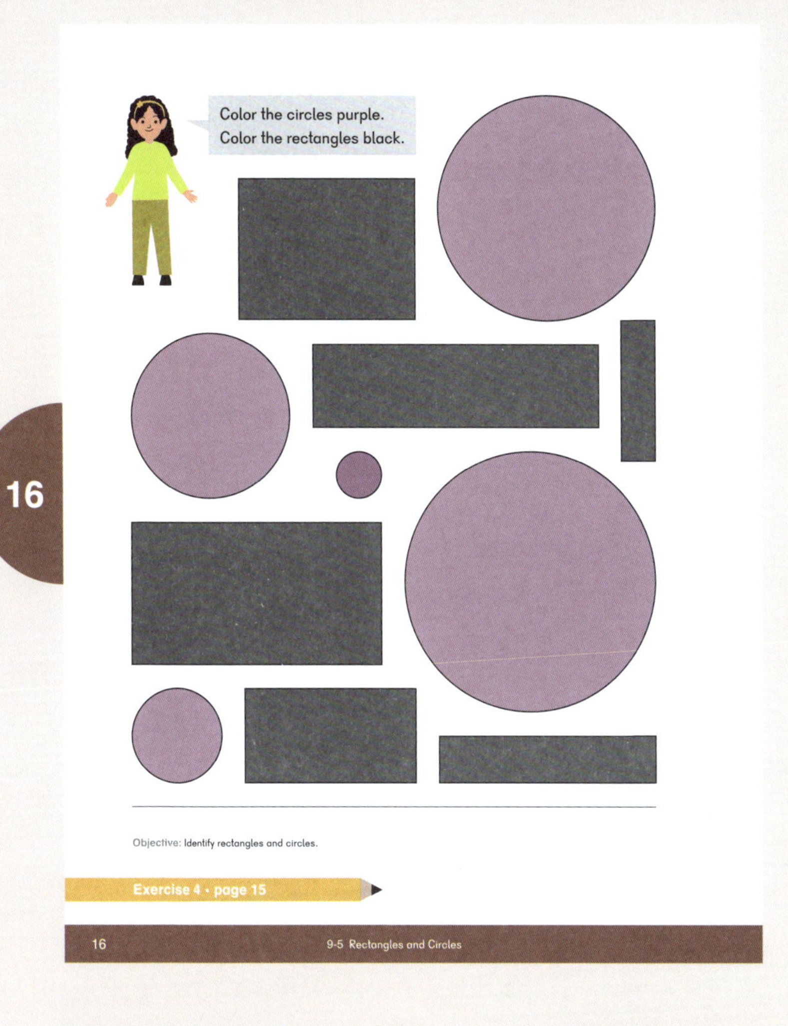

Whole Group Play

Hula Hoop Roll/Cylinder Roll: If you are able to go outside, give each pair of students a hula hoop and have them roll the hoop to each other. Alternatively, have students roll cylinders back and forth to each other and discuss why some roll farther than others.

Small Group Center Play

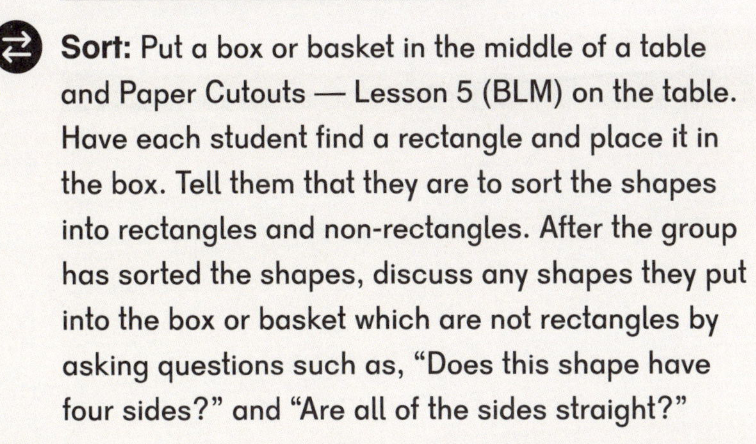

Sort: Put a box or basket in the middle of a table and Paper Cutouts — Lesson 5 (BLM) on the table. Have each student find a rectangle and place it in the box. Tell them that they are to sort the shapes into rectangles and non-rectangles. After the group has sorted the shapes, discuss any shapes they put into the box or basket which are not rectangles by asking questions such as, "Does this shape have four sides?" and "Are all of the sides straight?"

Building: Have students build structures using up to ten cylinders and identify the number of building pieces used.

Art: Have students decorate their papier-mâché spheres made during the last lesson.

Have students trace around the face of a cylinder, then color the circle.

Extend Learn

Draw a Rectangle: Give each student a piece of art paper, a straight edge, and a crayon. Using a straight edge, draw a rectangle on the board or project the drawing of a rectangle. Have students draw a rectangle using a straight edge for the sides, then color it.

Exercise 4 • page 15

Objectives

- Recognize squares as parts of cubes.
- Identify squares.

Lesson Materials

- Plastic or foam cubes of different sizes, 1 per student
- Poster board or large piece of art paper
- Ribbon
- Paint
- Geoboards and rubber bands
- Square cut out of paper
- Optional snack: square-shaped crackers

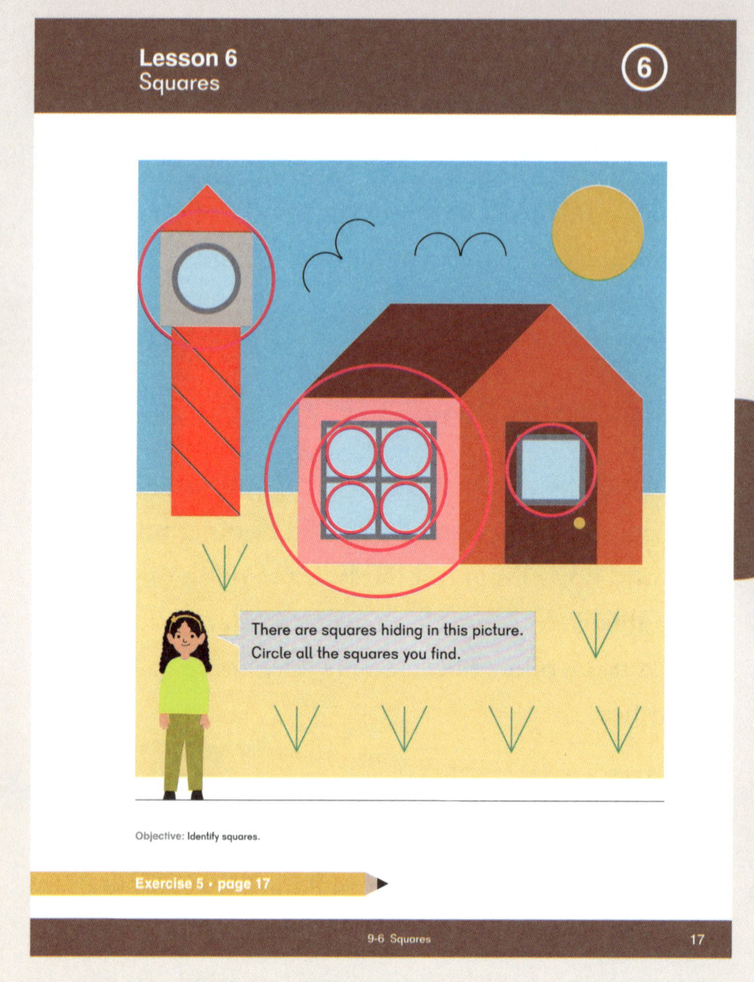

17

Explore

Give each student a cube. Rub one face of a cube and have students do the same. Ask students the name of the shape on the face of a cube. If necessary, introduce the word "square." Ask:

- How many sides does a square have?
- Are the sides curvy or straight?
- What do you notice about the lengths of the sides of a square?
- How many corners does a square have?
- Are some sides longer than others?

Use a ribbon cut to the same length as the square's sides and have students hold the ribbon to the sides of the square to prove they are all the same length.

Be sure they understand that a square has four corners, four sides that are all the same length, and that the sides must be straight.

Dip a face of a cube into paint and transfer the paint to a poster board. Have students discuss the relationship between a cube and a square. Be sure that they understand that a cube is solid and a square is flat.

Hold up a square and ask students to name the shape. Turn the square 45 degrees and ask them to name the shape again.

Learn

Have students look at page 17 and circle all of the squares.

Whole Group Play

You're Such a Square!: Put students of similar heights in groups of four. Have them use their bodies on the floor to create squares.

Sing "Draw a Square" (lyrics on page 16). Teach students the song. Use your finger to draw a square in the air and have students do the same. Emphasize that all sides of a square are of equal length by having students sing that part of the song louder than the rest.

Small Group Center Play

Build Squares: Have students create as many squares as possible on a geoboard.

Trace and Color Squares: Have students trace around the faces of cubes, then color the squares.

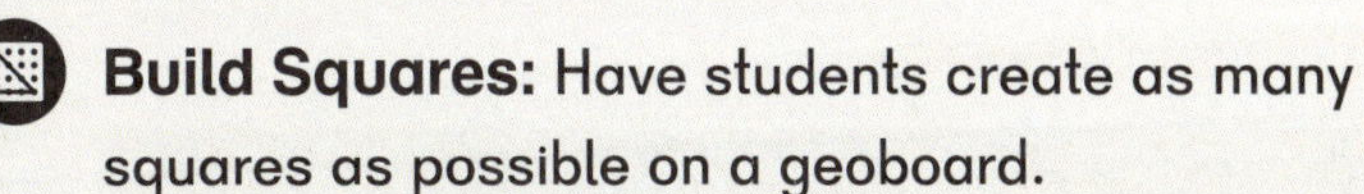

Extend Explore

Cut the Square: To help develop visualization skills, hold up a square and have students imagine that you cut the square right in half through the middle. Use your fingers to show where the cut would be. Have them tell you what shapes you made (rectangles).

If students are unable to visualize, fold the square and cut it showing the two rectangles.

Objective

- Identify triangles.

Lesson Materials

- Right triangle cut out of paper
- Paper Cutouts — Lesson 7 (BLM), 1 of each per pair of students
- Box or basket,1 per pair of students
- Rounded toothpicks or coffee stirrers, 3 per student
- Modeling clay or play dough
- Geoboards and rubber bands
- Paper cutouts of triangles, circles, squares, and rectangles
- Googly eyes
- Optional snack: wonton triangles pre-cut into triangles, olive oil, and Parmesan cheese

Explore

Hold up a right triangle and ask students to name the shape. If necessary, introduce the word "triangle." Ask:

- How many sides does a triangle have?
- Are the sides straight or curved?
- How many points does a triangle have?

Turn the triangle 45 degrees and 90 degrees and ask them to name the shape. Be sure they understand that a triangle has 3 straight sides and 3 points.

Give each pair of students a set of Paper Cutouts — Lesson 7 (BLM) and a box or basket. Have them sort their shapes into triangles (in the container) and non-triangles.

After a few minutes, have students hold up their non-triangle shapes. Discuss any shapes they hold up that are triangles by asking questions such as:

- Does this shape have three sides?
- Are all of the sides straight?
- Does it have 3 points?

Learn

Have students look at page 18 and cross out the appropriate shapes.

Whole Group Play

Cook Up Some Triangles!: Pre-cut wonton wrappers diagonally to create triangles. Lay them flat on parchment paper on top of a baking sheet. Have students brush olive oil on the triangles and sprinkle Parmesan cheese on them. Bake at 375 degrees for 5 to 8 minutes. Enjoy!

Triangle Hokey Pokey: Give each student paper cutouts of triangles, circles, squares, and rectangles. Change the lyrics of the song:

Put your triangle in.
Take your triangle out.
Put your triangle in,
and shake it all about.
You do the hokey pokey and you turn yourself around.
That's what it's all about.

Small Group Center Play

Make a Triangle: Provide rounded toothpicks or coffee stirrers (3 per student) and modeling clay or play dough. Teach students how to roll clay into small balls to poke with toothpicks to create triangles. Have them make triangles.

Triangle Fish: Provide paper cutouts of triangles of different sizes, googly eyes, and glue. Show students how to use a small triangle to make a tail and a large triangle to make the body of the fish. Have them glue an eye to the fish.

Exercise 6 • page 19

Extend Explore

Geoboard Triangles: Challenge students to make as many triangles as they can on a geoboard. None of the triangles can look the same.

To help develop visualization skills, hold up a rectangle and have students imagine that you cut the rectangle diagonally. Use your fingers to show where the cut would be. Have them tell you what shapes you made (triangles).

If students are unable to visualize, fold the rectangle and cut it showing the two triangles.

Objective

- Review squares, circles, rectangles, and triangles.

Lesson Materials

- Paper Cutouts — Lesson 8 (BLM)
- Paper cutouts of triangles, circles, squares, and rectangles
- Containers to sort into
- Pictures of a unicycle, bicycle, and tricycle
- Optional snack: carrot coins

Lesson 8
Squares, Circles, Rectangles, and Triangles — Part 1 (8)

It's a circle and you know it cause it's round.
It's a circle and you know it cause it's round.
It rolls very well.
It's curved you can tell.
It's a circle and you know it cause it's round.
It's a triangle and you know it has 3 sides.
It's a triangle and you know it has 3 sides.
It's also got 3 corners.
Which are pointy, kind of sharp.
It's a triangle and you know it has 3 sides.
It's a rectangle and you know it has 4 sides.
It's a rectangle and you know it has 4 sides.
It's a face on a box.
Now you're clever like a fox.
It's a rectangle and you know it has 4 sides.
A square also has 4 sides.
A square also has 4 sides.
A cube's face is a square.
You can always find it there.
A square also has 4 sides.

Objective: Learn some attributes of squares, circles, rectangles, and triangles.

9-8 Squares, Circles, Rectangles, and Triangles — Part 1 19

Explore

Place paper cutouts of shapes from Paper Cutouts — Lesson 8 (BLM) around the classroom. Send students on a treasure hunt to bring one of each shape back to the group.

Have students bring the shapes they found to the group, name the shapes they brought, and tell how they knew those were the shapes. Encourage them to talk about numbers of sides, numbers of corners, and whether sides are straight or curved.

Learn

Place all the paper cutouts for this lesson on the floor in front of the students. As they sing the song, have them find and hold up paper cutouts of the shape they are singing about. Sing the song several times, telling students to find a different example of the shape each time they sing about it.

Have students look at page 19. Read Mei's speech bubble to them and have them sing the song again, using their fingers to trace around the outside of the shapes as they sing about each.

Read Sofia's speech bubble on page 20 to students and have them complete the task. Discuss the kite. Some students may call that shape a "diamond." Be sure they understand that it is a square, given its attributes.

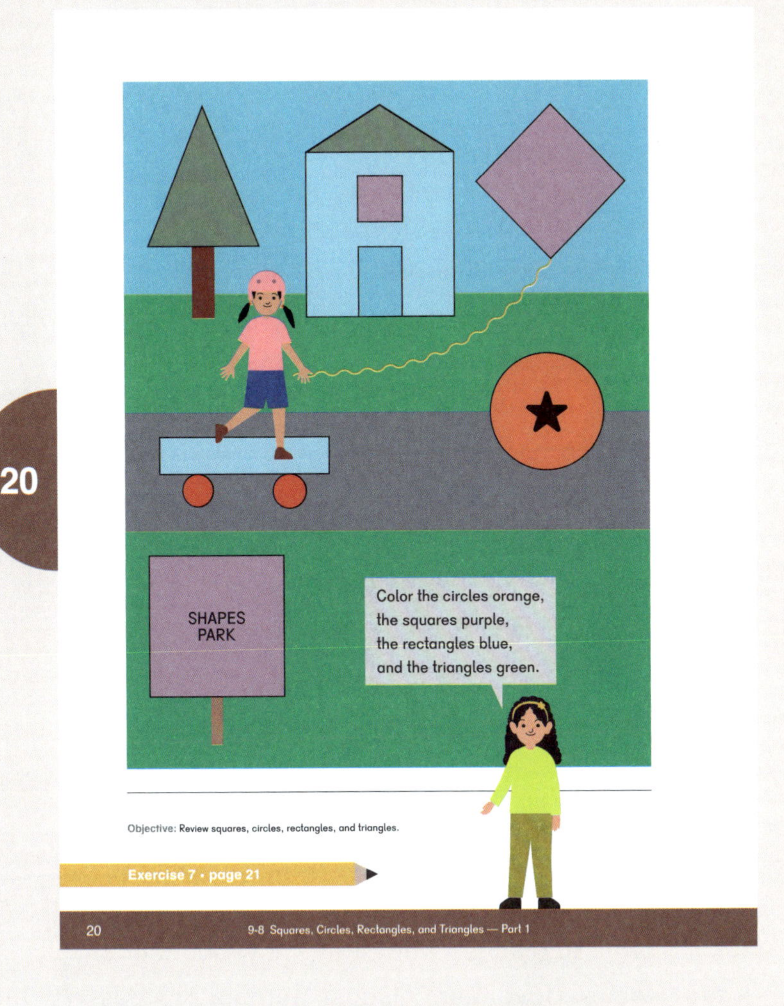

Whole Group Play

Have students work in pairs. Students take turns using their fingers to draw either a circle, rectangle, or triangle on their partner's back. The partner not drawing must guess the shape that was drawn.

Small Group Center Play

Sort: Set out four sorting containers and paper cutouts of circles, squares, rectangles, and triangles of different colors and sizes. Have students sort the cutouts in as many ways as they can.

Shape Pictures: Set out art paper, paper cutouts, and glue. Tell students to use the materials to create certain pictures that you suggest. One example would be to create a house. A cutout of a triangle could be the roof, and cutouts of rectangles could be the building, the door, and windows. A small circle could be used as a doorknob. Other ideas include a car, snowman, and fruit tree.

Exercise 7 • page 21

Extend Learn

1, 2, 3 Wheels: Ask students if they have ever ridden on a tricycle. Why are the wheels round? Would square wheels work? Show students pictures of a unicycle, bicycle, and tricycle. Discuss the level of difficulty of balancing on each one and why that is the case.

Objective

- Identify examples and counterexamples of squares, circles, rectangles, and triangles.

Lesson Materials

- Wooden or plastic cubes, rectangular prisms, and cylinders
- Paper Cutouts — Lesson 9 (BLM), 1 cutout per student
- Pipe cleaner bubble wands, pre-made (directions on page 16)
- Bubble soap (dish soap and water)
- Pipe cleaners
- Pictures of art masterpieces
- Large chevron
- Optional snack: banana coins and square cheese slices

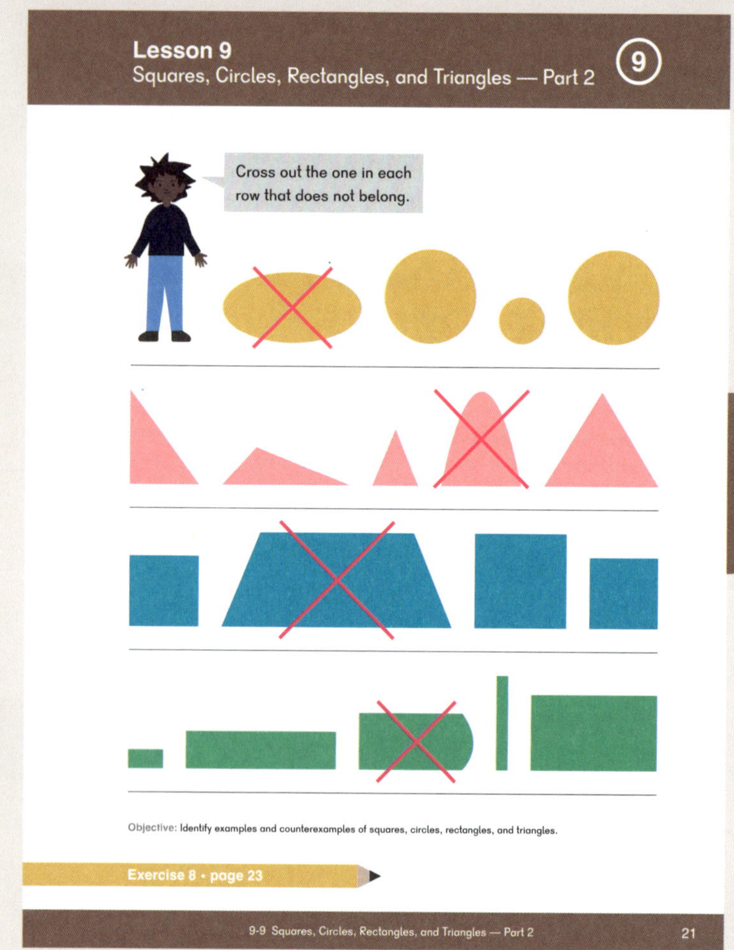

Explore

Ask students to discuss what they know about each of the shapes. Have them discuss number of sides, number of corners, and whether sides are straight or curved.

Hold up a cutout of a trapezoid from Paper Cutouts — Lesson 9 (BLM), and ask students if it is a square. Students should notice that it is not a square because the lengths of its sides are not the same.

Hold up examples and counterexamples of each shape and have students discuss. If the shape you hold up is a counterexample, have students tell why that shape is not an example of the shapes they are learning about.

Learn

Have students look at page 21. Read Dion's directions and have students complete the task.

Whole Group Play

Provide pipe cleaner bubble wands and bubble soap. Name a shape and have students dip their wands in the soap and blow bubbles. Repeat with different shapes.

Small Group Center Play

Building: Provide pipe cleaners. Have students bend pipe cleaners to create triangles and rectangles. Discuss the shapes they create.

Art: Show students pictures of art masterpieces, such as "The Snail" by Henri Matisse, "The Starry Night" by Vincent van Gogh, or "Sunflowers" by Vincent van Gogh. Have students discuss familiar shapes in the paintings. Then provide paints and art paper and have them create their own masterpieces.

Exercise 8 • page 23

Extend Explore

I Used to be a Triangle: Post a picture of a large chevron. Have students talk about the difference between a triangle and a chevron. Then have them record imaginative stories about what might have happened to a triangle in order for it to become a chevron.

Objective

- Practice concepts introduced in this chapter.

Lesson Materials

- Bubble wands from previous lesson
- Solids
- Paper cutouts of shapes (used in previous lessons from this chapter)
- Craft sticks
- Optional snack: snacks in the shapes of cubes, cylinders, and spheres, such as, cheese cubes, gelatin cubes, marshmallows, turkey and cheese pinwheels, grapes, and puff cereal

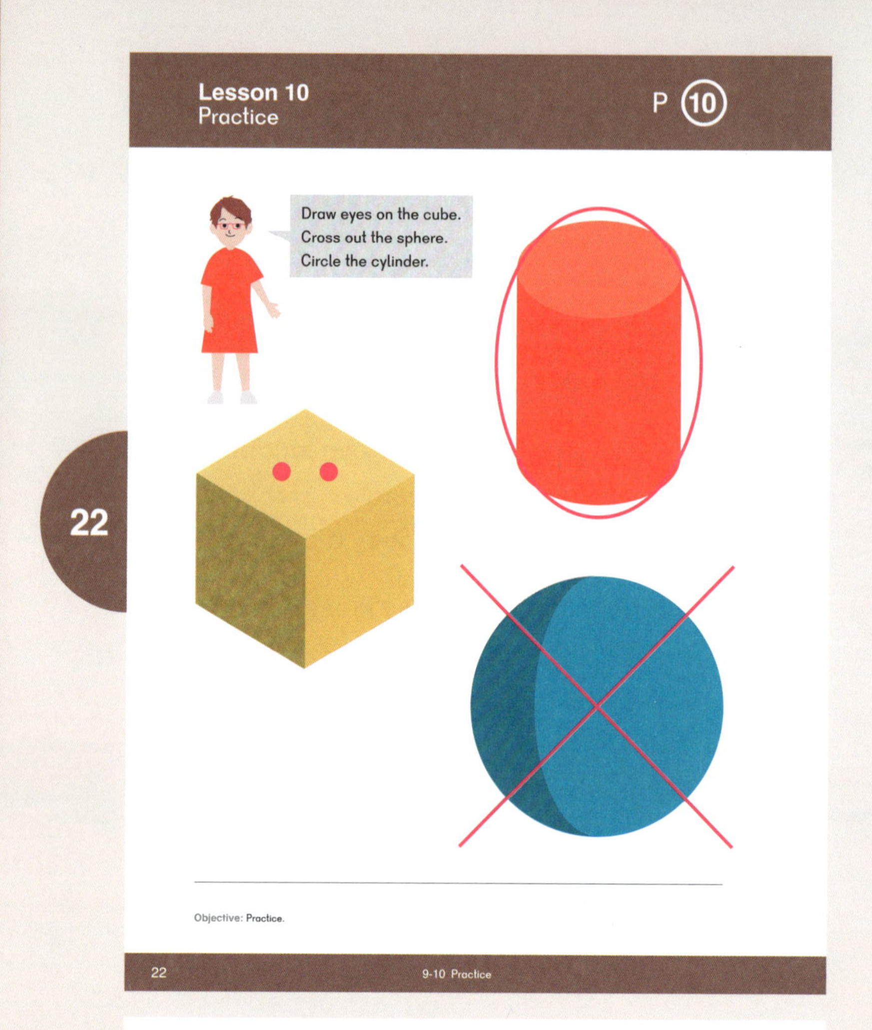

For the **Practice**, read the directions and speech bubbles on each page and have students complete the tasks.

Whole Group Play

Use bubble wands and blow bubbles.

Sing the songs learned in this chapter.

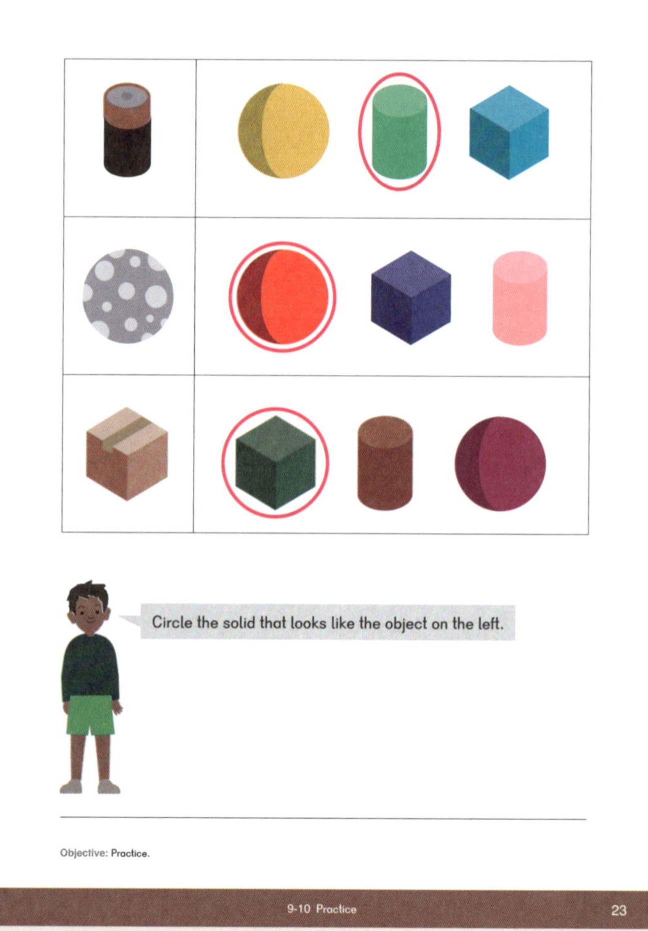

Small Group Center Play

Sort: Provide numerous examples of solids and shapes, and have students sort them in different ways.

Building: Have students create a structure and then make up a story about who lives or works in it. They can either draw pictures to illustrate their stories or record them.

Art: Have students trace around the faces of cubes and cylinders to create squares and circles. Have them describe the positions of the shapes on their papers using positional words.

Exercise 9 • page 25

Extend Play

Crafting Shapes: Provide craft sticks and glue. Have students create squares and triangles. When the glue dries, have them decorate their shapes using paint, glitter, crayons, etc.

Exercise 1 • pages 9–10

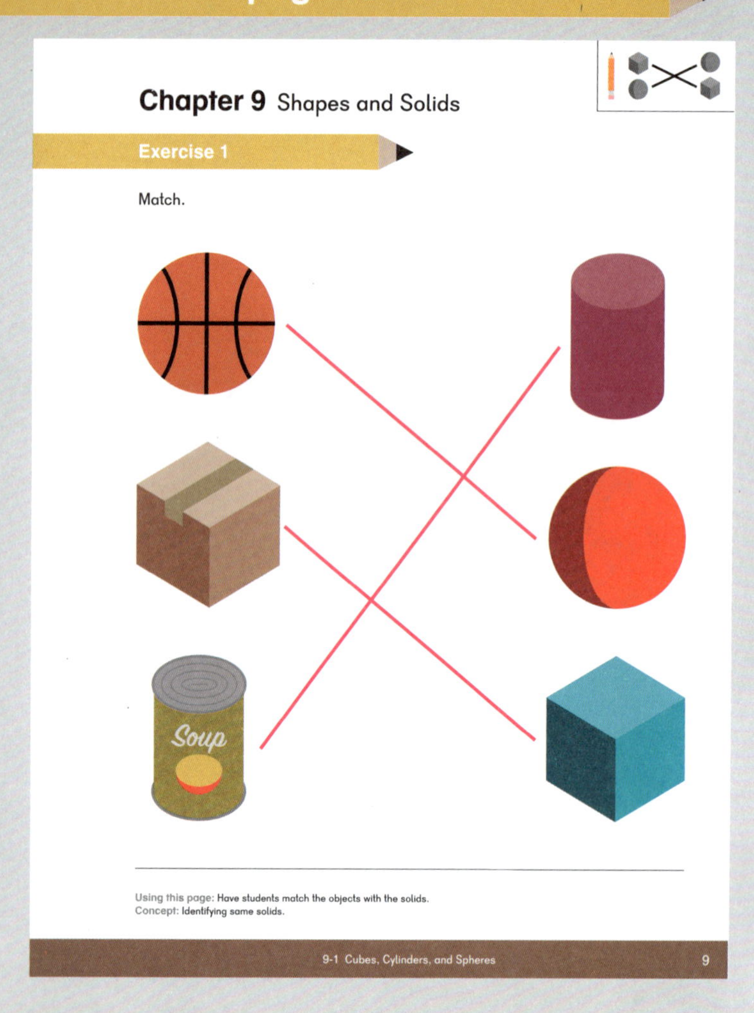

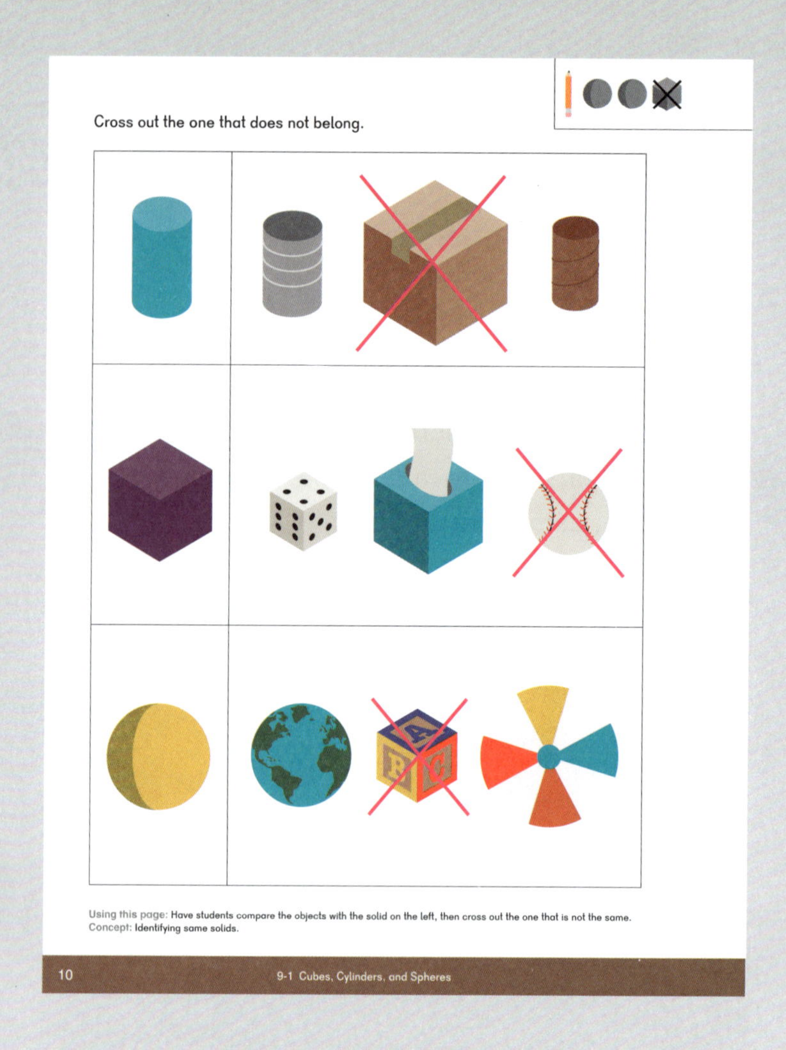

Exercise 2 • pages 11–12

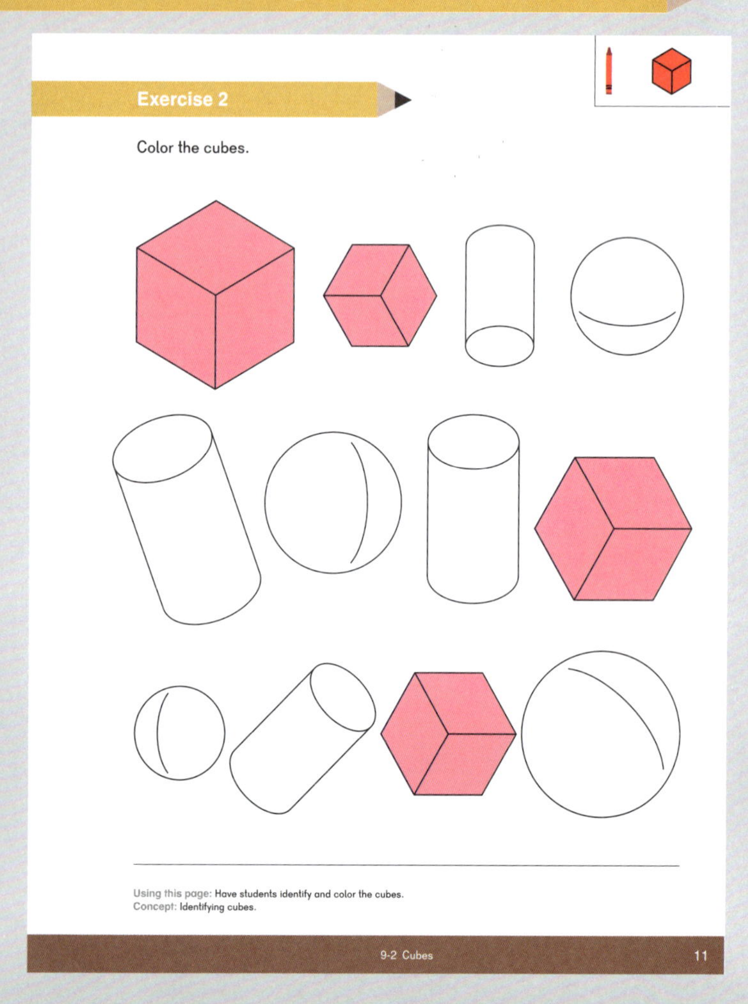

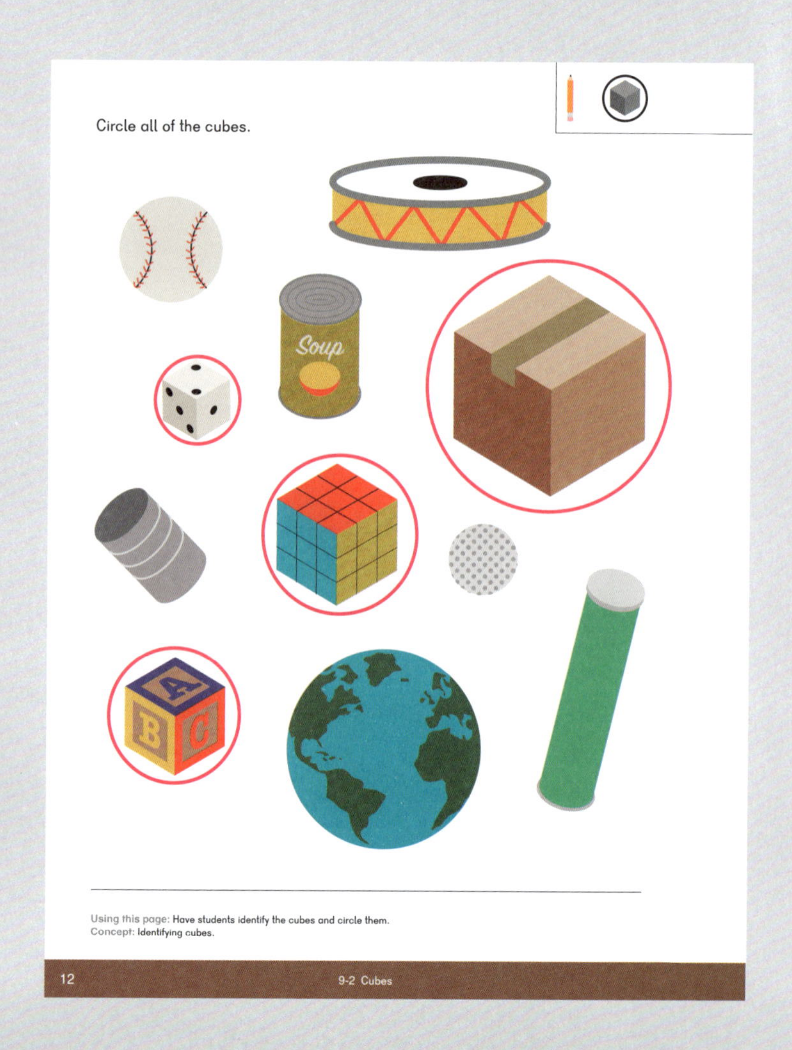

Exercise 3 • pages 13–14

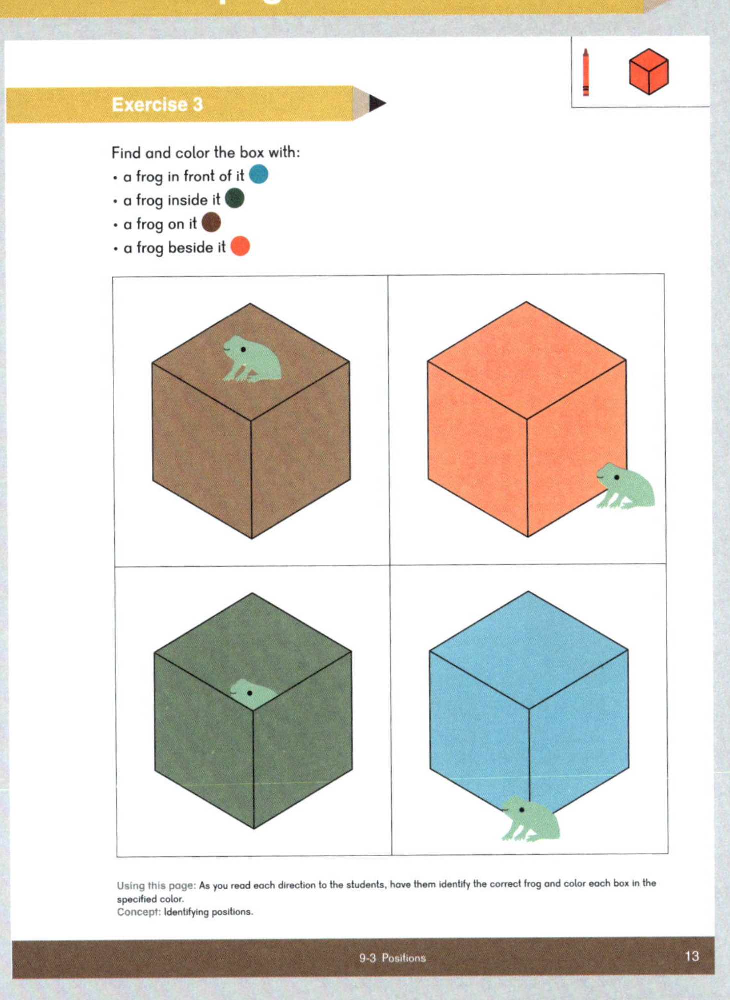

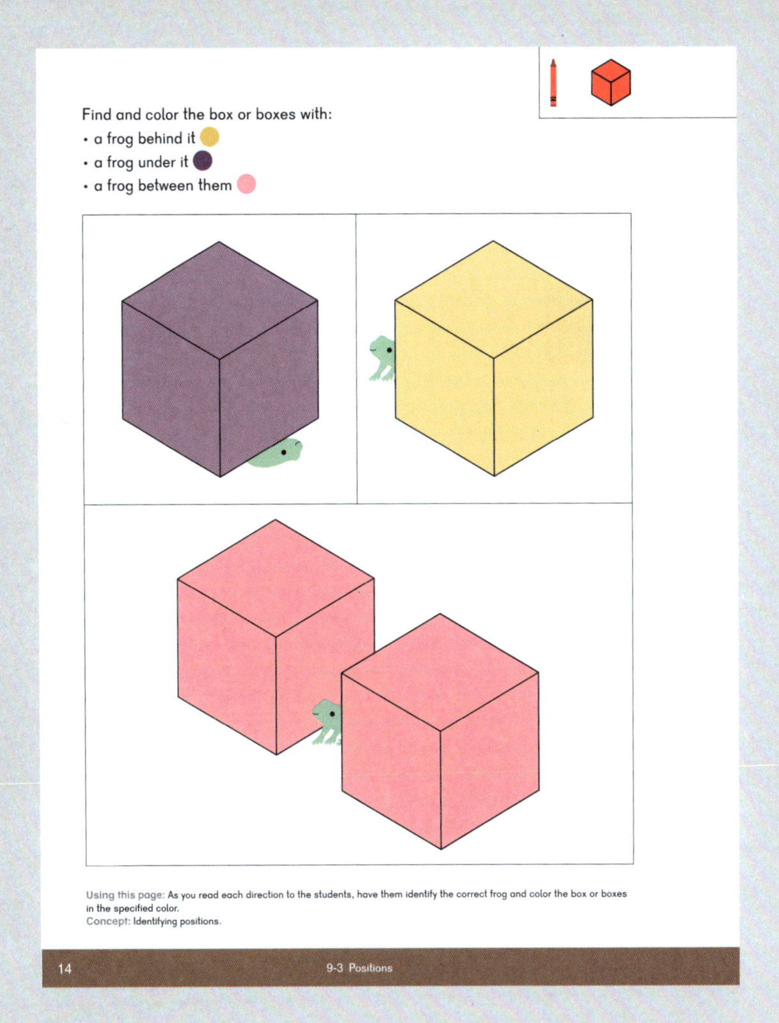

Exercise 4 • pages 15–16

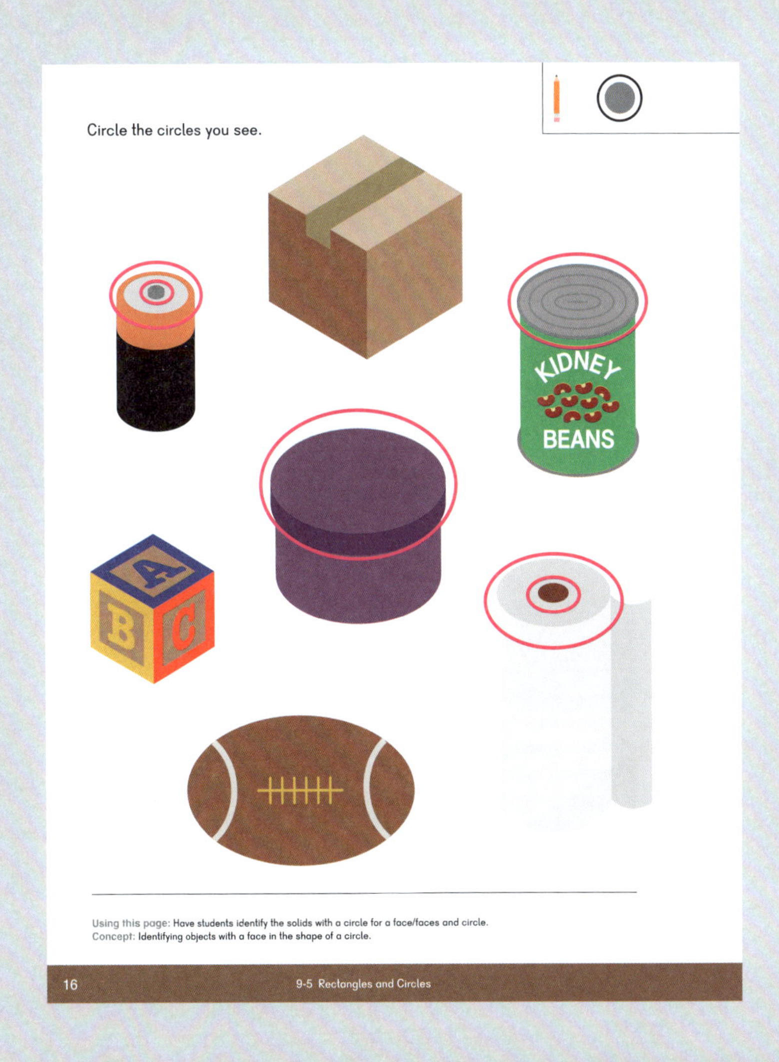

Exercise 5 • pages 17–18

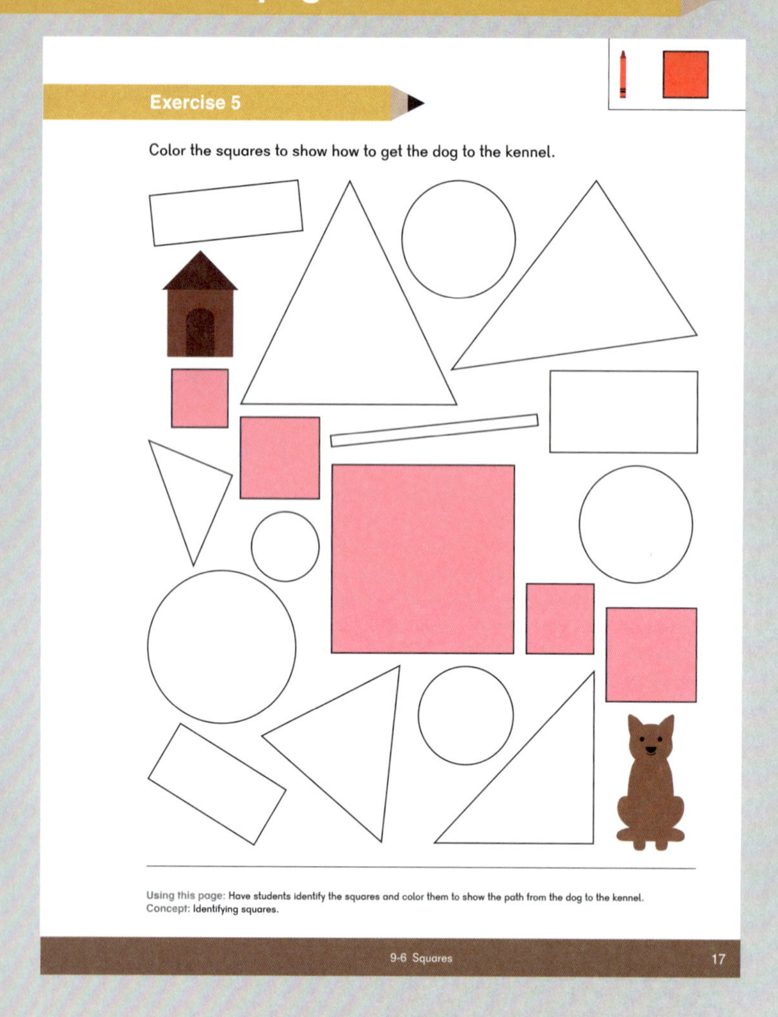

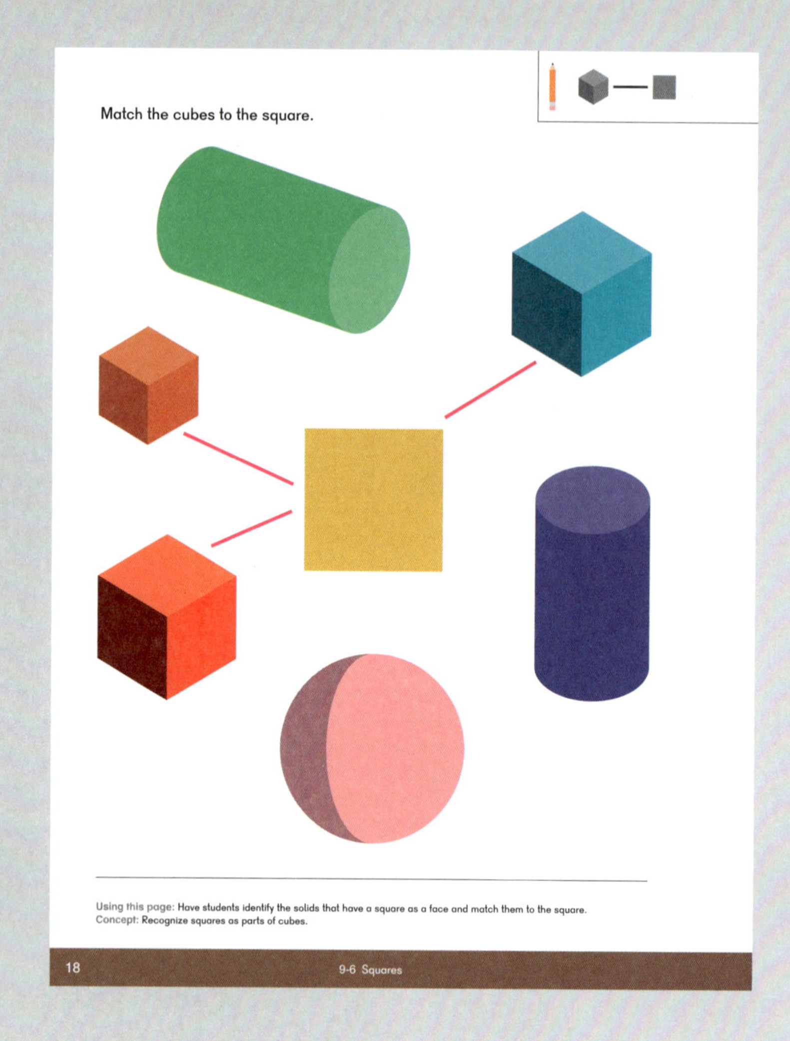

Exercise 6 • pages 19–20

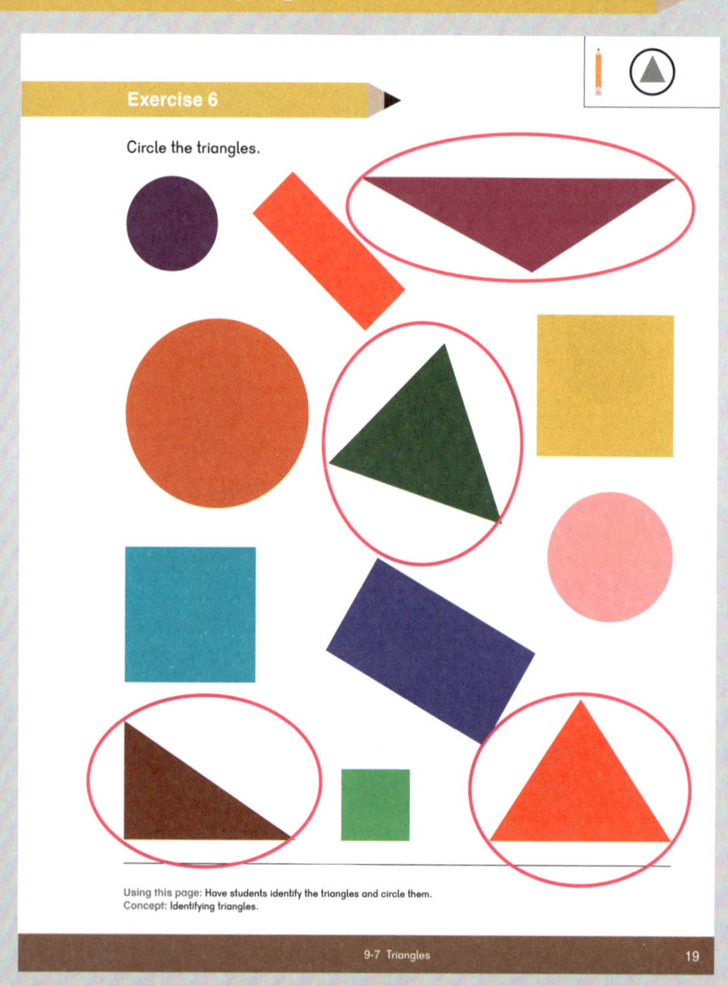

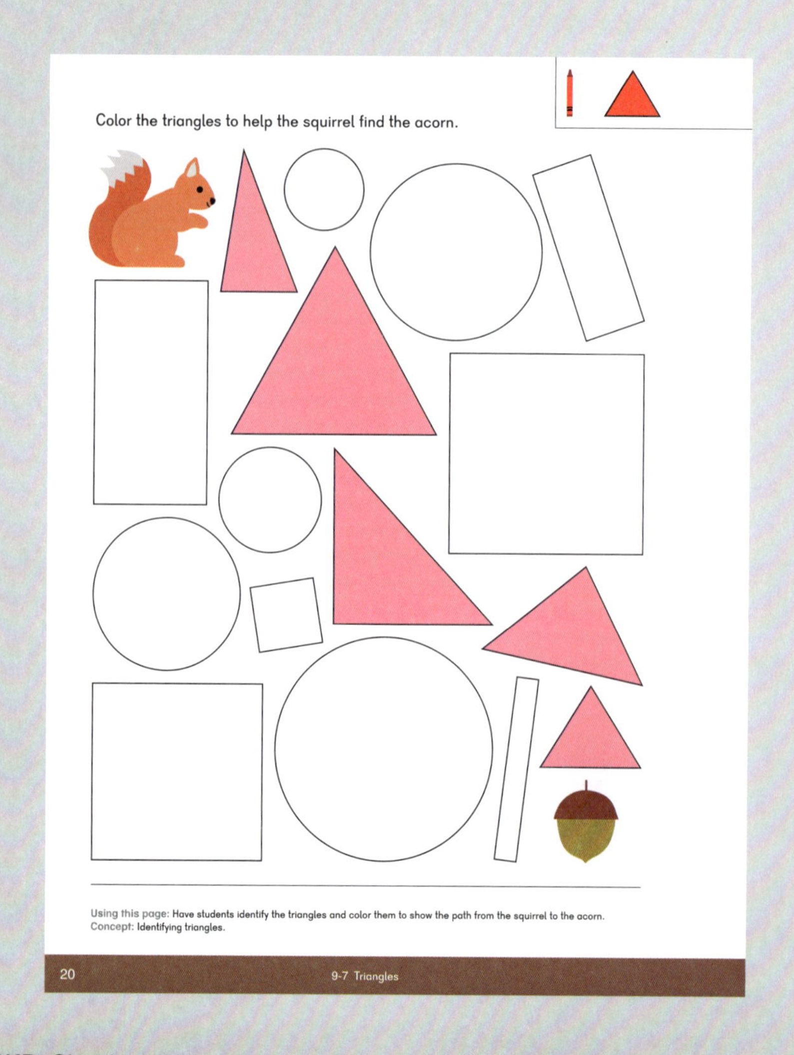

Exercise 7 • pages 21–22

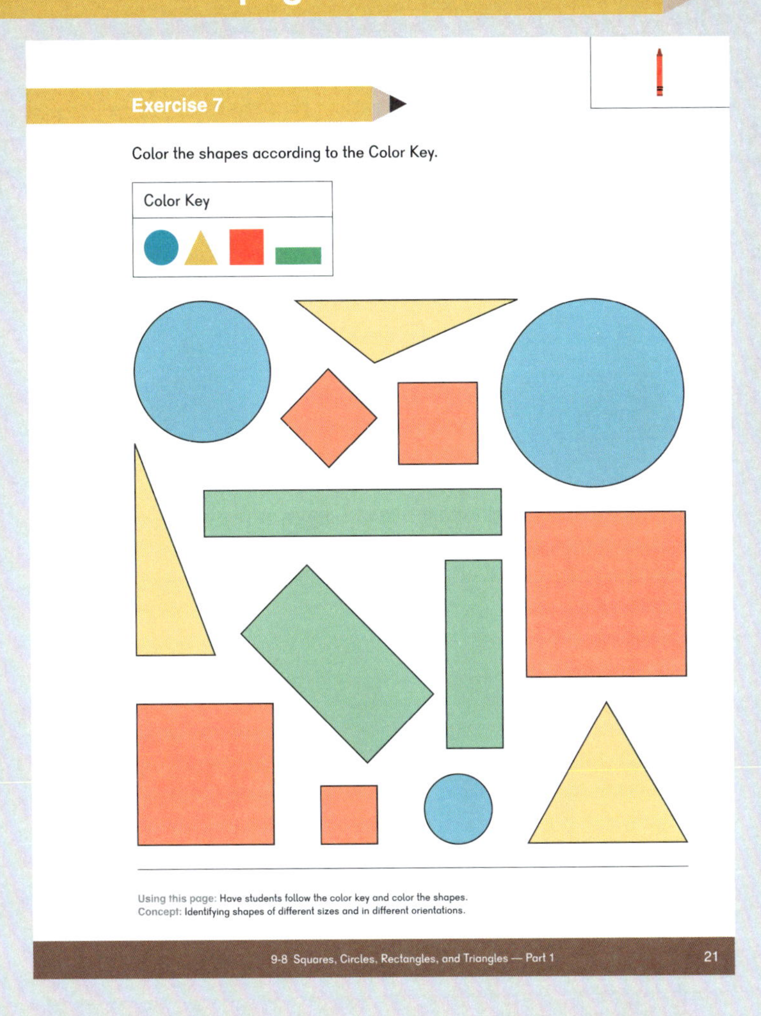
Exercise 7

Color the shapes according to the Color Key.

Color Key

Using this page: Have students follow the color key and color the shapes.
Concept: Identifying shapes of different sizes and in different orientations.

9-8 Squares, Circles, Rectangles, and Triangles — Part 1 21

Color the shapes according to the Color Key.

Color Key

Using this page: Have students identify the shapes in the picture and color them according to the color key.
Concept: Identifying shapes of different sizes and in different orientations.

22 9-8 Squares, Circles, Rectangles, and Triangles — Part 1

Exercise 8 • pages 23–24

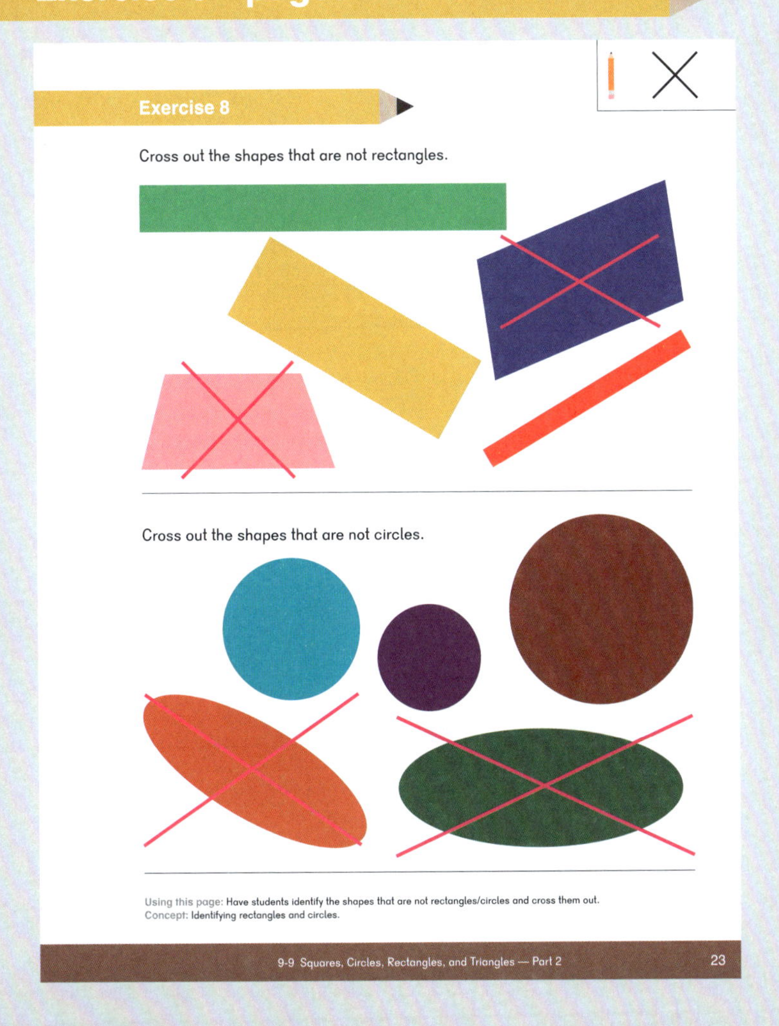
Exercise 8

Cross out the shapes that are not rectangles.

Cross out the shapes that are not circles.

Using this page: Have students identify the shapes that are not rectangles/circles and cross them out.
Concept: Identifying rectangles and circles.

9-9 Squares, Circles, Rectangles, and Triangles — Part 2 23

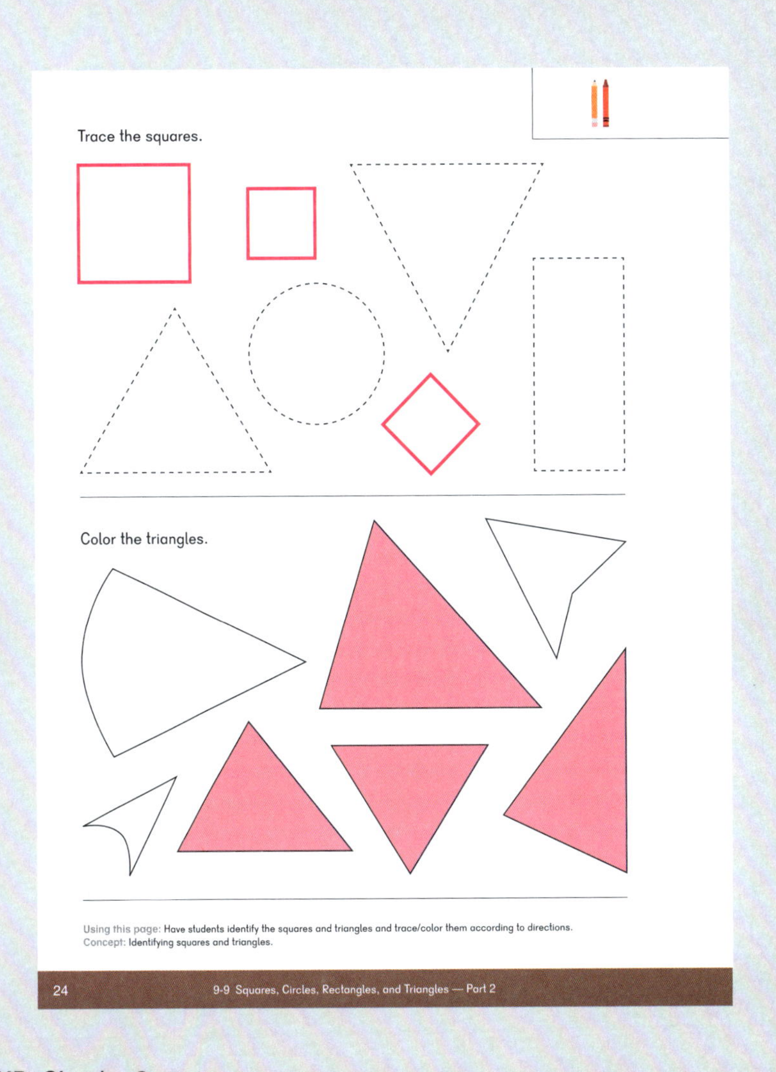
Trace the squares.

Color the triangles.

Using this page: Have students identify the squares and triangles and trace/color them according to directions.
Concept: Identifying squares and triangles.

24 9-9 Squares, Circles, Rectangles, and Triangles — Part 2

Exercise 9 • pages 25–26

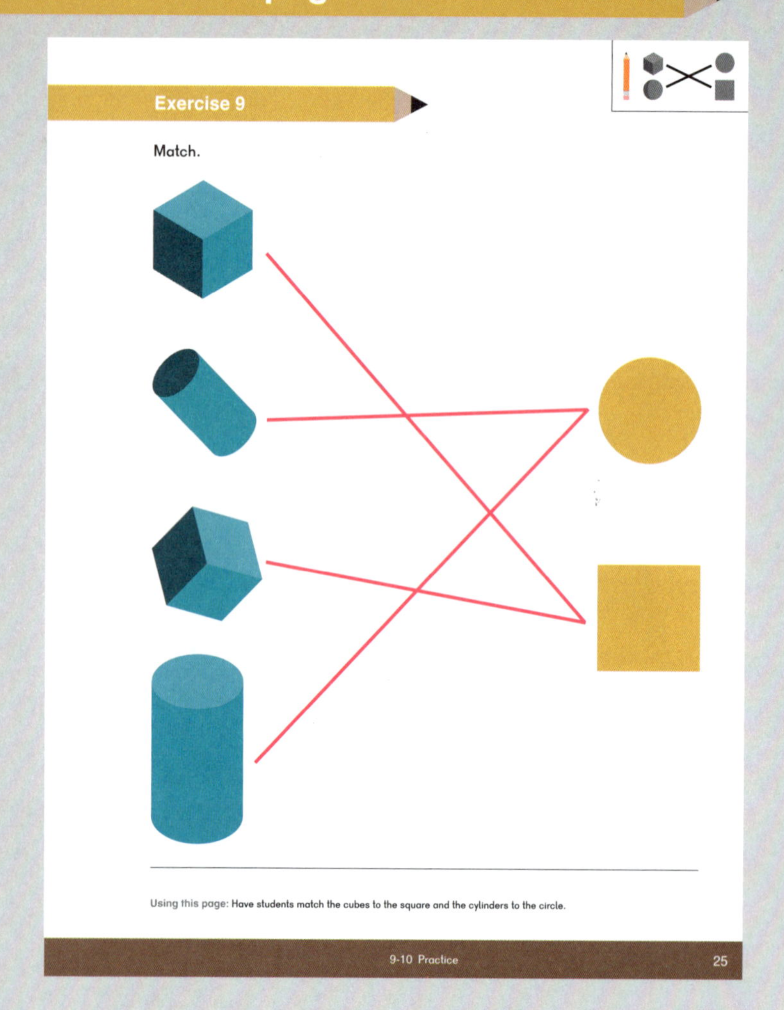

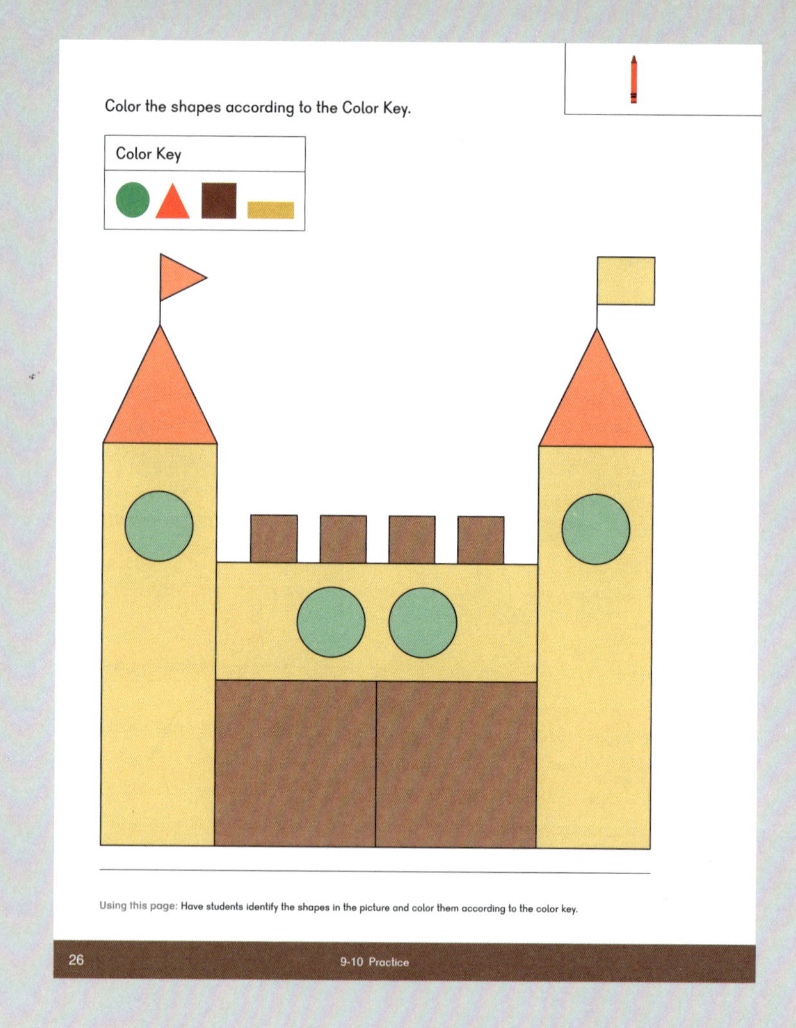

Suggested number of class periods: 5–6

	Lesson	Page	Resources	Objectives
	Chapter Opener	p. 47	TB: p. 25	
1	Match Objects	p. 48	TB: p. 26 WB: p. 27	Match one object to one object.
2	Which Set Has More?	p. 51	TB: p. 30 WB: p. 31	Compare two sets of objects to find which set has more.
3	Which Set has Fewer?	p. 53	TB: p. 33 WB: p. 33	Compare sets of objects to find sets that have fewer.
4	More or Fewer?	p. 56	TB: p. 36 WB: p. 35	Review comparing sets of objects to determine which has more or fewer.
5	Practice	p. 58	TB: p. 39 WB: p. 37	Practice concepts introduced in this chapter.
	Workbook Solutions	p. 60		

Chapter Vocabulary

- Compare
- More
- Fewer
- Same number

Notes

Comparing quantities of objects allows Pre-Kindergarten students to begin to make sense of numbers. This concept is introduced in this chapter by comparing sets of objects. Students will begin by matching objects one-to-one to see if there is one ____ for every ____. They will move on to comparing sets of objects that have different numbers in order to answer which set has more and which set has fewer.

Key Points

Similar to how some students find subtraction more difficult than addition, for some students, "fewer" is a more difficult concept than "more."

While students may not have heard "fewer," they have likely heard "less." Connect fewer and less. Grammatically, "fewer" is used for items that are individually counted, while "less" is used for aggregate quantities. We say, "fewer cups of water." or, "less water."

At this age, students need not correctly associate "fewer" with countable items and "less" with uncounted quantities, but teachers should practice modeling the correct language.

When working with manipulatives in this chapter, be sure that students begin by lining up the sets one-to-one in order to compare the quantities in each. If necessary, use empty egg cartons and/or ten-frame cards to help with this skill.

Materials

- Chairs
- Construction paper
- Counters
- Cups
- Dishes
- Doll or stuffed toy
- Finger paints
- Linking cubes
- Napkins
- Paper plates
- Pipe cleaners
- Plastic bags
- Play dough
- Play food
- Red and blue crayons
- Small counters
- Small objects for sorting
- Spoons
- Stamps
- Strips of paper
- Tissue paper

Note: Materials for Activities will be listed in detail in each lesson.

Blackline Masters

- Bicycle/Tricycle Cards
- Hey Diddle Diddle Cards
- Picture Ten-frame Cards
- Ten-frame Cards

Storybooks

- *The Very Hungry Caterpillar* by Eric Carle
- *More, Fewer, Less* by Tana Hoban

Optional Snacks

- Rice cake moons and milk
- Cheerios
- Raisins
- Banana slices
- Celery sticks
- Yogurt or cream cheese
- Cherry tomatoes

Letters Home

- Chapter 10 Letter

Notes

Lesson Materials

- 2 or more types of play food to sort
- Other small objects to sort

Explore

Have students help you sort play food into groups in different ways. Each time a sort is completed, have students explain the reason for the sort. Then ask them what they can say about how many of each type of food there is. Do not arrange the groups in any order.

Give each pair of students small objects to sort in different ways and have them repeat what was done with the play food.

Extend Explore

More or the Same: Give each pair of students two types of small objects. Ask them how they can decide if there are more of one type of object or the same number of each.

Chapter 10

Compare Sets

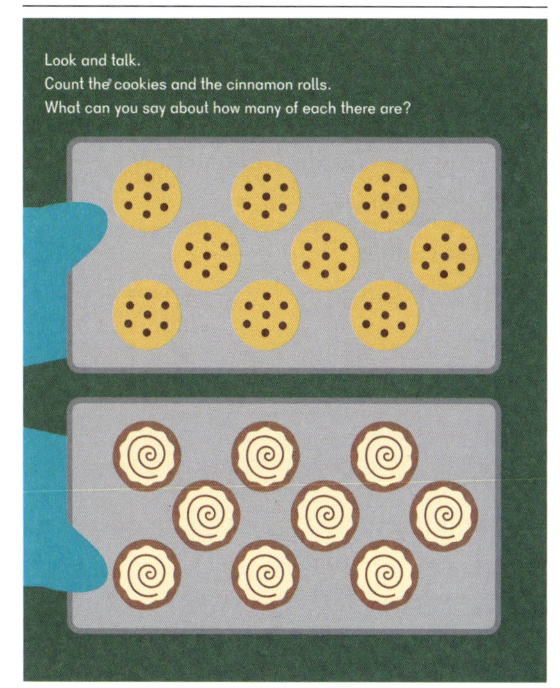

25

25

Objective

- Match one object to one object.

Lesson Materials

- Small objects, 2 types, 7 of each type per pair of students
- Play dishes and spoons
- Hey Diddle Diddle Cards (BLM)
- Small counters
- Finger paints
- Strips of paper
- Napkins
- Spoons
- Cups
- Dishes
- Optional snack: rice cake moons and milk

Explore

Give each pair of students 7 small objects of one type and 7 of another. Ask them how they can determine if there are more of one type of object or if there is the same number of each. Observe the problem solving strategies students use to answer the question.

Learn

Show students a pile of small dishes and a pile of spoons, 8 of each. Ask them what you should do if you want to find out if there is a dish for each spoon. If necessary, count the dishes and place them in front of you in a row. Then count the spoons and line them up against the dishes to show that there is one spoon for each dish. Ask, "Is there a dish for each spoon?" Say, and have students repeat, "There is the same number of dishes and spoons."

26

Lesson 1
Match Objects ①

Hey diddle diddle,
The cat and the fiddle,
The cow jumped over the moon.
The little dog laughed
To see such sport,
And the dish ran away with the spoon.

Objective: Match one object to one object.

26 10-1 Match Objects

Read "Hey Diddle Diddle" and have students put a finger on each picture as it is mentioned. Ask them:

- Is there a fiddle for each cat?
- Is there a cow for each moon?
- Is there a dish for each spoon?

Read Alex's direction and have students use a finger first, then a pencil to draw lines. Ask students if there is a cat for each fiddle. Say, and have them repeat, "There is the same number of cats and fiddles."

Read Mei's question on page 28. Observe how many students are able to answer the question by subitizing. Encourage students to subitize whenever possible. Then have students use a pencil to match the spoons and dishes. Ask students if there is the same number of dishes and spoons.

Read Dion's speech bubble on page 29.

Ask students how page 29 is different from the previous pages. They may notice that there is less space for a line to be drawn and that it may take more thought to decide which dog to match with which bowl.

Have students use a pencil to match dogs to bowls.

Ask students if there is a bowl for each dog.

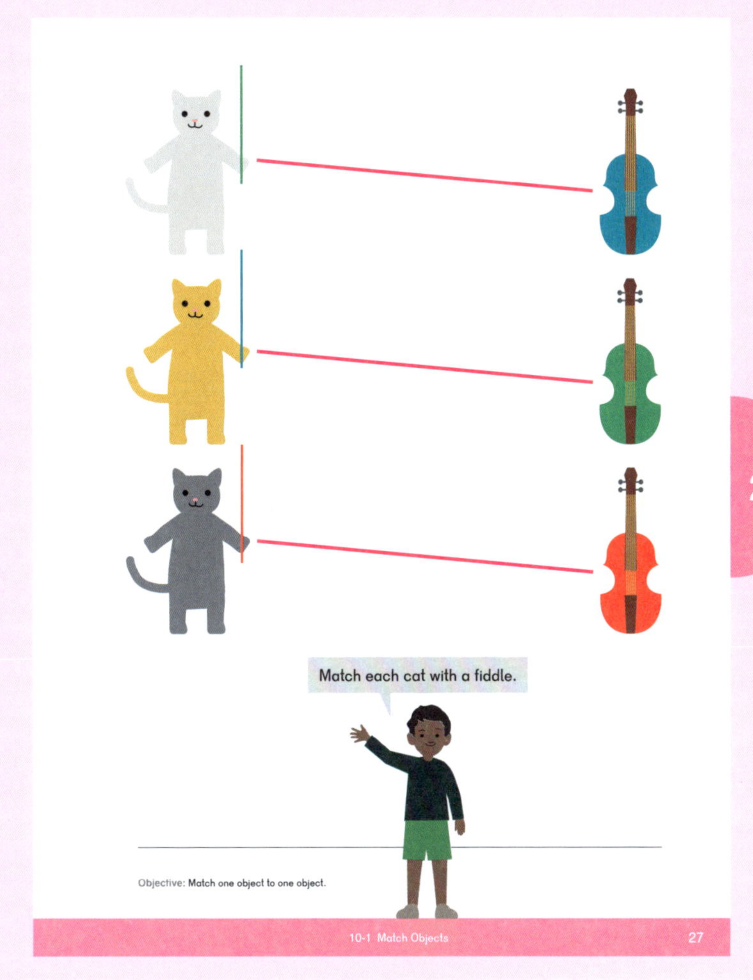

Whole Group Play

Have students line up in two lines, holding hands with a partner. Have them notice that there is the same number of people in each line.

Play "London Bridge is Falling Down" (VR). Have the first two students in line hold their hands up high to form an arch. The other students walk under the arch as everyone sings the song. When the word "lady" at the end of the verse is sung, the students forming the arch drop their arms and capture the student on the "bridge" at that time. Continue until all students have had a chance to form the arch. If the class is large, break students up into two different "bridges" so that there will be less waiting.

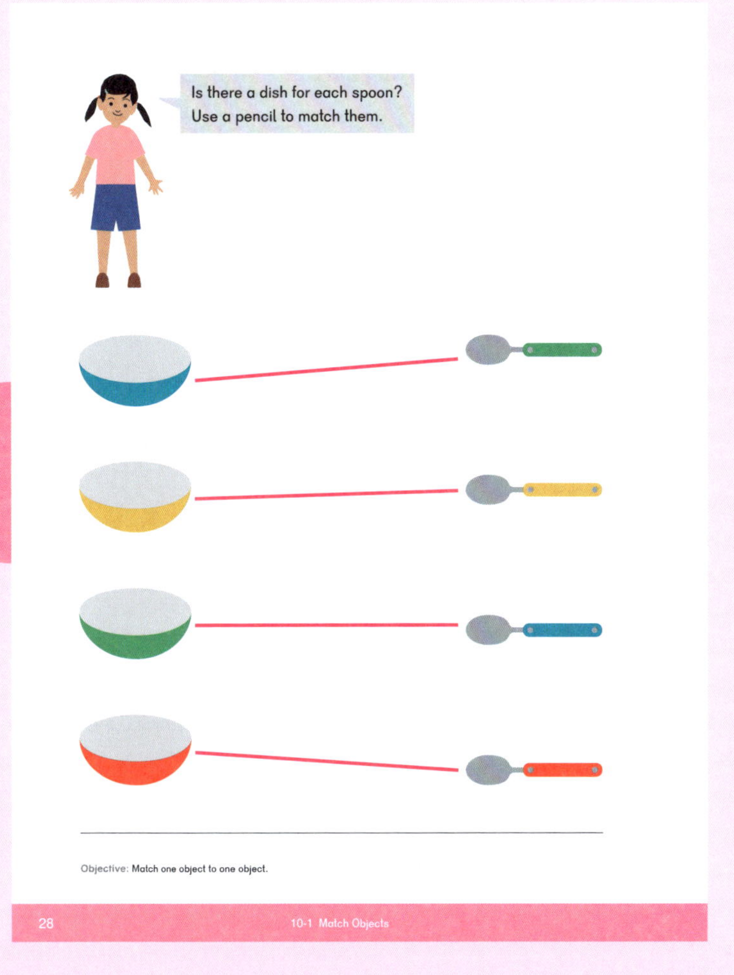

Small Group Center Play

Counting: Provide Hey Diddle Diddle Cards (BLM). Students pick a card and put the number of counters on the card to match the number of pictures.

Dramatic Play: Read "Hey Diddle Diddle" again and have students act it out.

Thumb Print Art: Dip your thumb in finger paint and make a row of 6 thumbprints on a strip of paper. Have students use strips of paper to show the same number of thumbprints. Allow them to turn their thumbprints into a masterpiece.

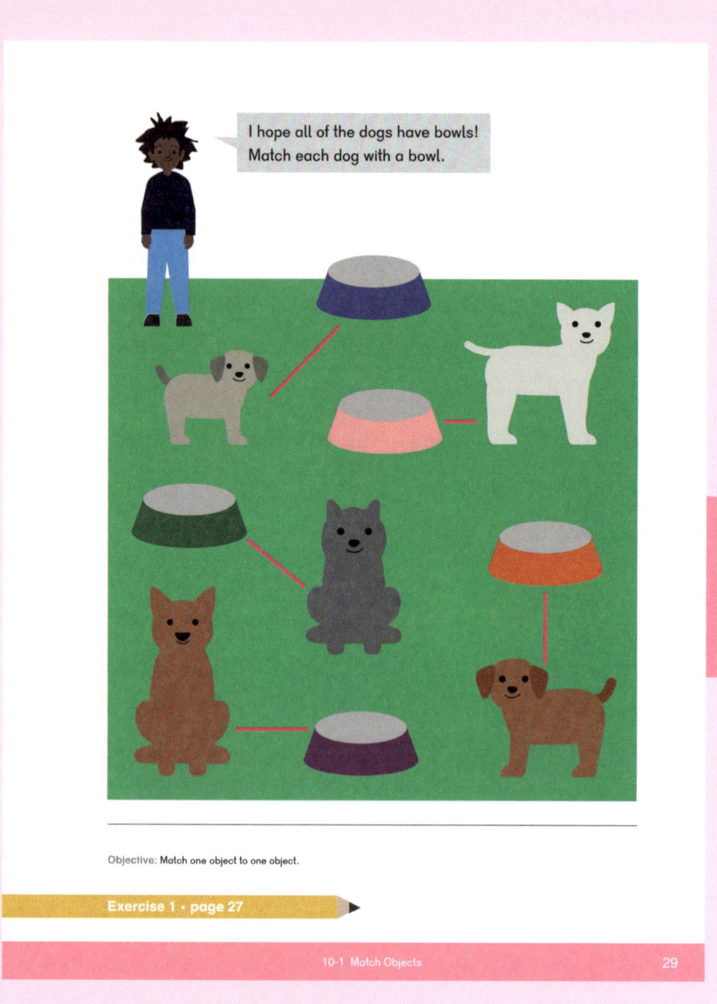

29

Exercise 1 • page 27

Extend Learn

Set the Table: Have students set a table for snack. Tell them that there must be one napkin, one spoon, one cup, and one dish for each student at the table.

Let's Rhyme: Read "Hey Diddle Diddle" to the students again. Have students think of words that rhyme with fiddle (middle, riddle, griddle) and moon (dune, June, noon, soon, tune), and say what the rhyming words mean.

Objective

- Compare two sets of objects to find which set has more.

Lesson Materials

- Bags of mixed red and blue crayons, less than 5 of each, 1 bag per pair of students
- Chairs set up for musical chairs, starting with 1 chair per student
- Picture Ten-frame Cards (BLM) 1 to 10
- Ten-frame Cards (BLM) 1 to 10
- Tissue paper or construction paper of 2 colors
- Optional snack: 5 banana slices and 6 raisins — Have students determine which snack there is more of and eat the raisins first

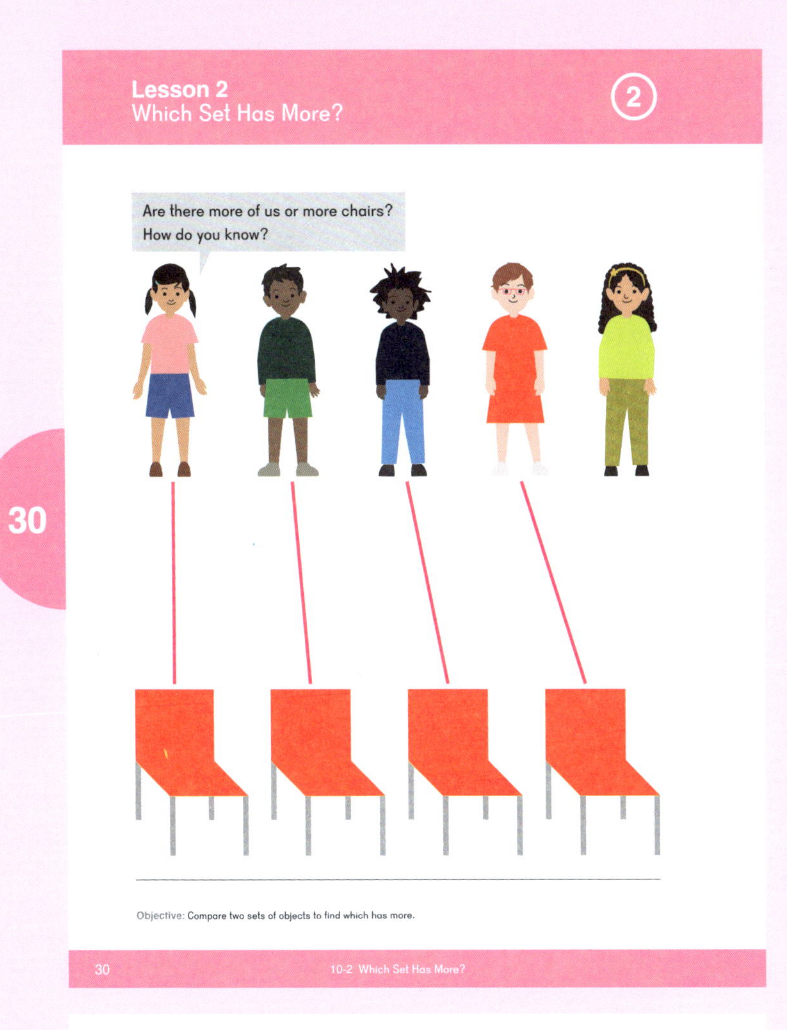

Explore

Give each pair of students a bag of red and blue crayons. Ask students, "Whose bag had more red crayons? How do you know? Whose bag had fewer red crayons? How do you know?" Repeat for blue.

Have students compare their bag with another pair's bag of crayons using the words "more" and "fewer."

Observe students' strategies for comparison. How many students count the number of objects in one set and then the number of objects in the other set? Do students line up the objects in order to compare? Have students discuss their strategies.

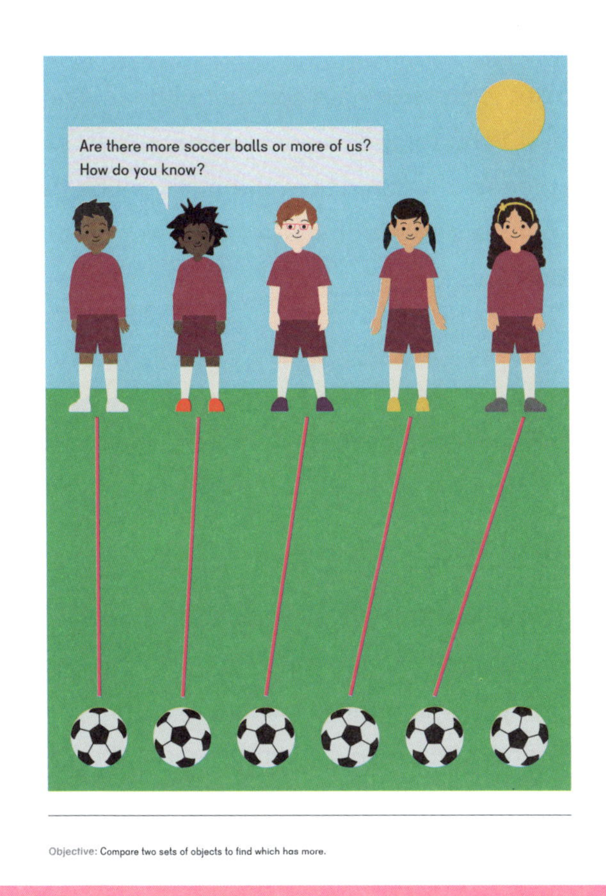

Learn

Have students look at page 30 and tell if there are more friends or more chairs and how they know.

Have students look at page 31 and answer Dion's questions.

Have students look at page 32. Read Emma's speech bubble. Have students discuss their methods of finding the answer.

Whole Group Play

Have students play musical chairs, beginning with an equal number of chairs and students.

With each round, have students comment on if there are more chairs or more students and fewer chairs or fewer students.

Small Group Center Play

Ten-Frame Flash: Hold up two Ten-frame Cards (BLM) at a time and have students point to the card that represents the greater number.

Counting: Provide Picture Ten-frame Cards (BLM) of chairs and students. Students choose a card showing chairs and a card showing students and tell which there are more of, which there are fewer of, or if there are the same number of each.

Collage: Have students create collages using two colors of tissue paper, construction paper, etc., up to 10 pieces of each. Have them tell which color has more pieces and which color has fewer pieces on their collages.

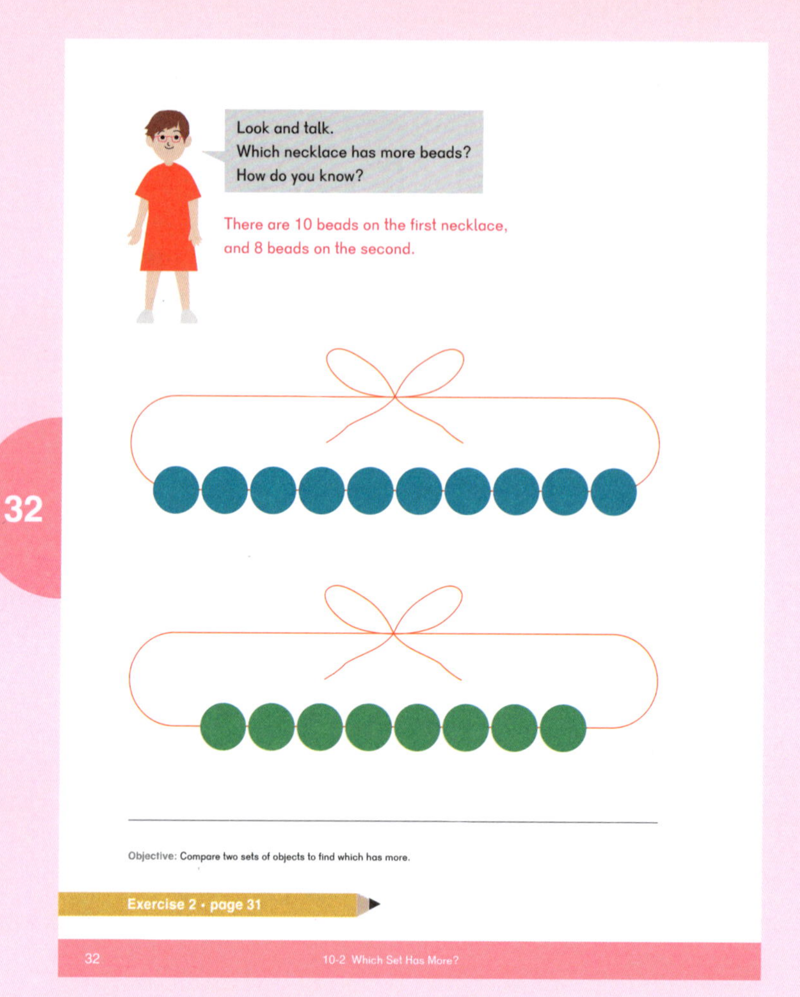

Exercise 2 • page 31

Extend Play

Ten-frame Flash with a Twist: Hold up 2 Ten-frame Cards (BLM) at a time for a few seconds. Have students tell which card shows a greater number by identifying the number shown on each, then saying which card shows more.

Objective

- Compare sets of objects to find sets that have fewer.

Lesson Materials

- Linking cubes, 12 per plastic bag, 1 bag per pair of students
- Counters
- Bicycle/Tricycle Cards (BLM)
- Stamps
- *More, Fewer, Less* by Tana Hoban
- Optional snack: banana slices — Have students pretend that each slice is a wheel for a tricycle. Have them count out the correct number of wheels and enjoy.

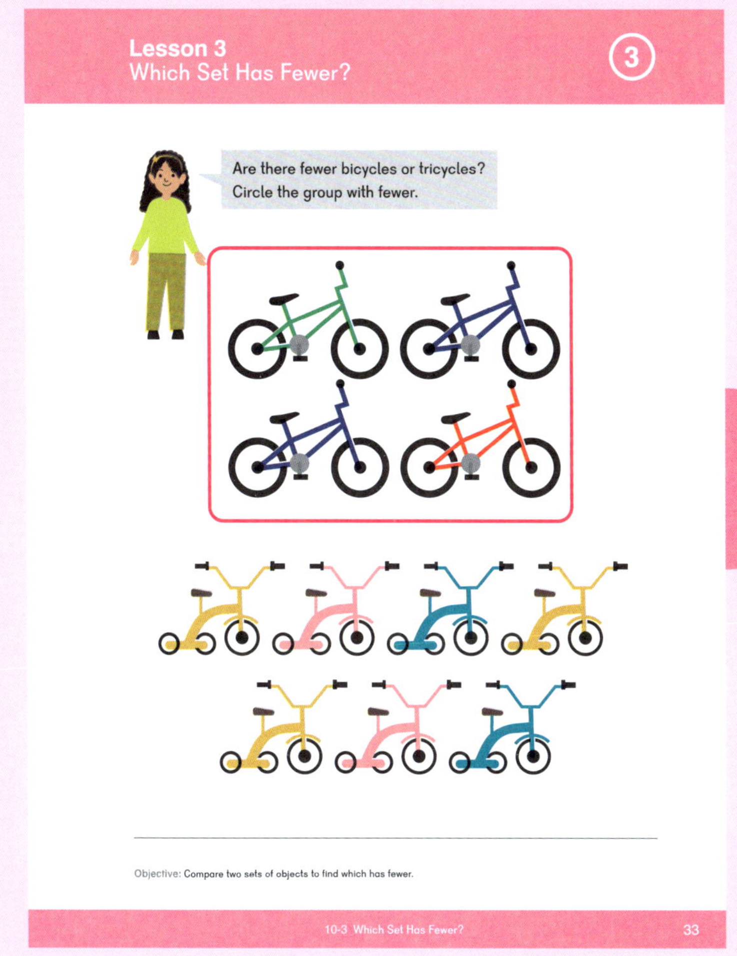

Explore

Give each pair of students a bag of linking cubes.

Have one student in each pair pick up a handful of linking cubes and set them on the floor.

Have the other student in each pair repeat, setting his or her linking cubes separate from those already on the floor.

Ask them to say something about the number of linking cubes in each set. Some pairs may have the same number in both sets. Other pairs should identify the set which has fewer linking cubes.

Learn

Lay out two rows of linking cubes, a row of 5 and a row of 3. Ask students which row has more cubes. Point to the other row and say, "This row has fewer cubes than the other row."

Have students look at page 33. Ask them if there are more bicycles or tricycles. Read them Sofia's speech bubble. Tell them that since there are more tricycles, there are fewer bicycles. Have them circle the bicycles.

Have students look at page 34 and tell how many legs the duck has and how many the cow has. Read Dion's speech bubble and have them tell which animal has fewer legs. Have them complete the task.

34

Which one has fewer legs?
Circle that one.

Objective: Compare two sets of objects to find which has fewer.

34 10-3 Which Set Has Fewer?

Have students look at page 35.

Read Mei's speech bubble and have them complete the task.

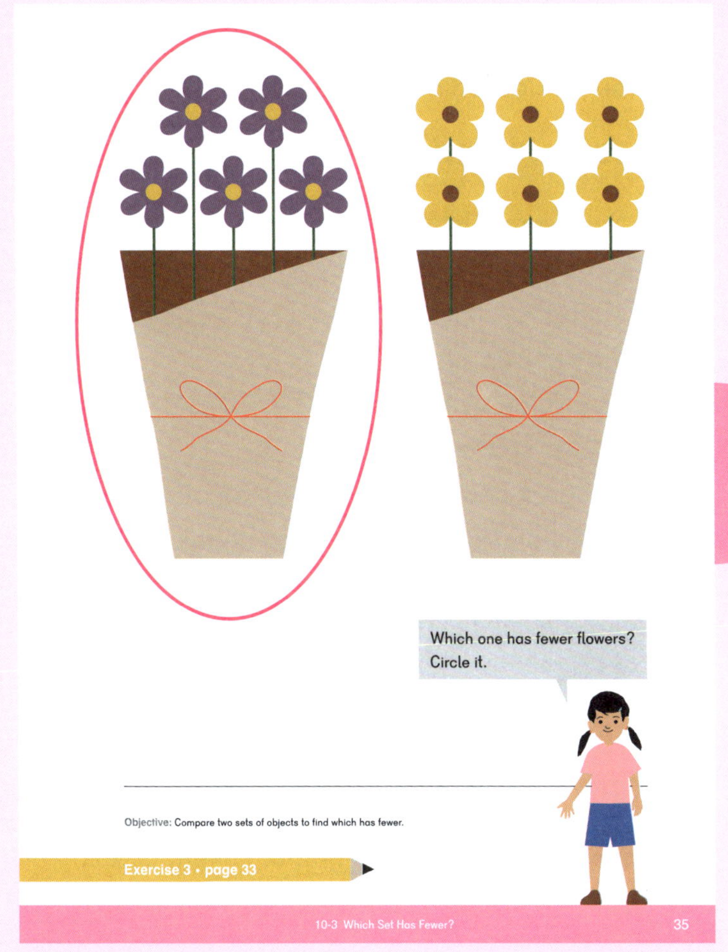

Whole Group Play

Read *More, Fewer, Less* by Tana Hoban. As this is a wordless picture book, discussing the pictures will help to deepen student understanding of mathematical concepts learned in Pre-K. For example, ask students to compare the number of objects on the pages. Have them point out sizes, positions, shapes, and textures.

Small Group Center Play

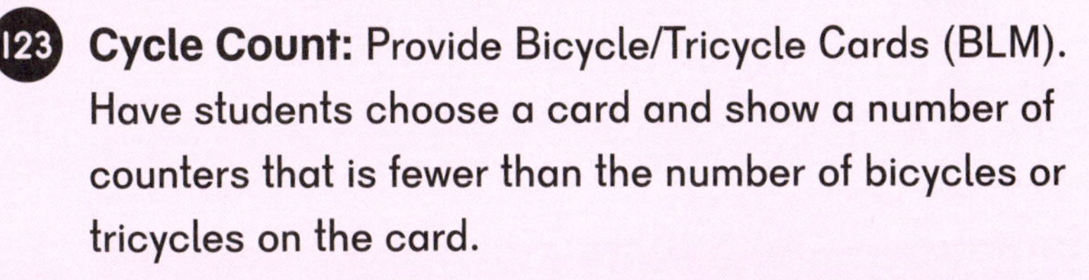

Cycle Count: Provide Bicycle/Tricycle Cards (BLM). Have students choose a card and show a number of counters that is fewer than the number of bicycles or tricycles on the card.

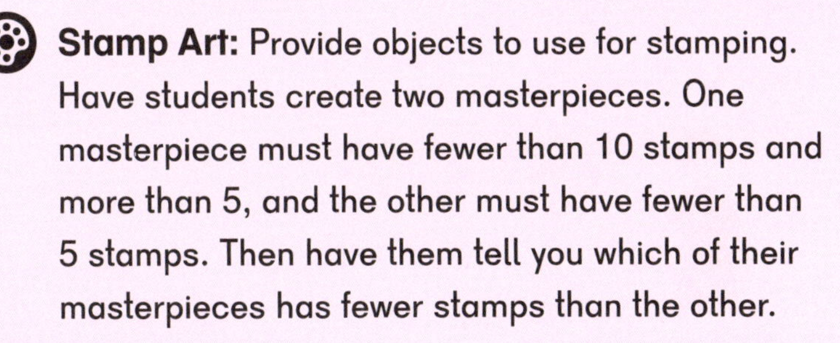

Stamp Art: Provide objects to use for stamping. Have students create two masterpieces. One masterpiece must have fewer than 10 stamps and more than 5, and the other must have fewer than 5 stamps. Then have them tell you which of their masterpieces has fewer stamps than the other.

Animal Actors: Have children choose to be either a two-legged animal or a four-legged animal and act like the animal. If space allows, have them race to a finish line as the animal.

Exercise 3 • page 33

Extend Learn

Solve It: Give each pair of students 14 linking cubes. Have them work together to solve this problem: "Sasha has 8 linking cubes. Meg has 6 linking cubes. How can they make their number of cubes the same?"

Objective

- Review comparing sets of objects to determine which has more or fewer.

Lesson Materials

- Small bags holding different numbers of linking cubes (up to 10), 1 bag per pair of students
- Play dough
- Ten-frame Cards (BLM) 0 to 10
- Doll or stuffed toy
- Optional snack: two types of small fruit such as grapes and banana slices — Have students make fruit skewers using more (fruit 1) and fewer (fruit 2).

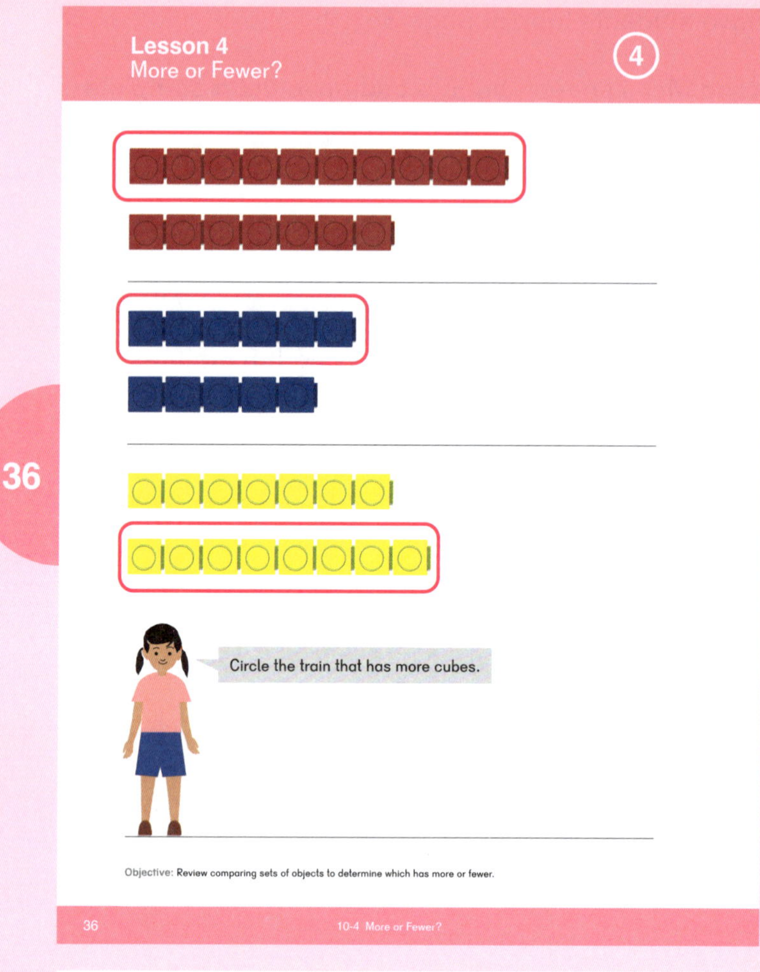

Explore

Give each pair of students a bag of linking cubes. Have students create towers with their cubes and then compare them with some other pairs' towers, using the words "more" and "fewer."

Learn

Hold up one of the towers from **Explore**. Count the cubes with the students. Then ask them what to do so that there will be more cubes in the tower. Add one linking cube and count again. Repeat with another linking cube. Lead students to understand that there can be many different situations where there can be more or fewer than a certain number of objects in a set.

Have students look at page 36. Read Mei's speech bubble to them and have them complete the task.

Have students look at page 37.

Read Emma's speech bubble and have them complete the task.

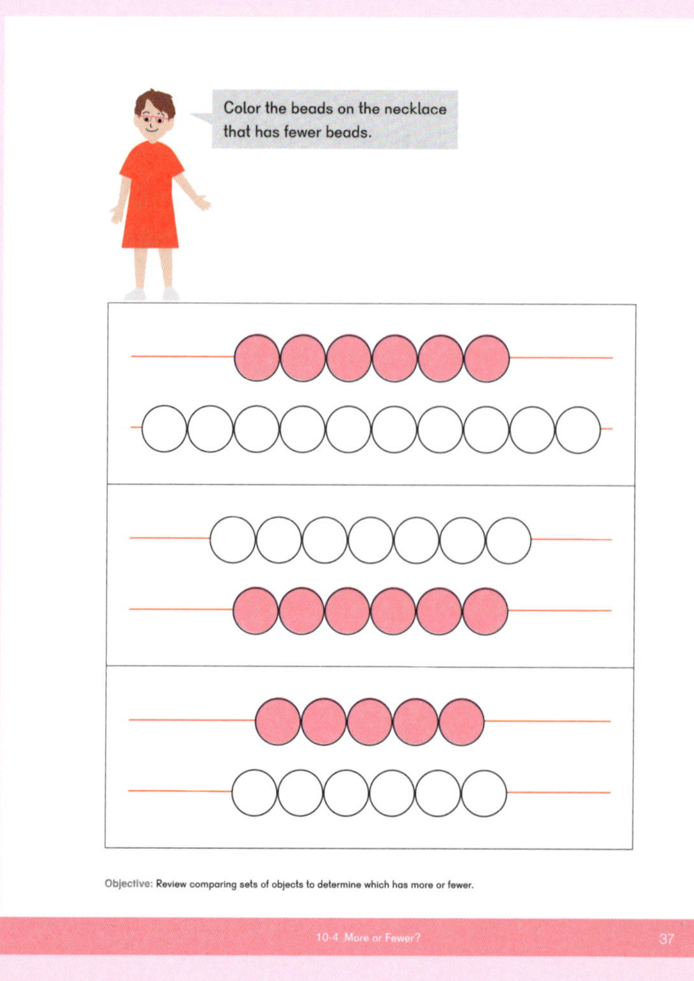

Have students look at page 38. Ask them to identify the creatures on the page. Read Mei's speech bubble to them. Give them some time to color the snakes in the second row. Ask if there are different ways to color the snakes in the second row so that more are colored than in the first row. Lead them to see that they could color either six or seven snakes.

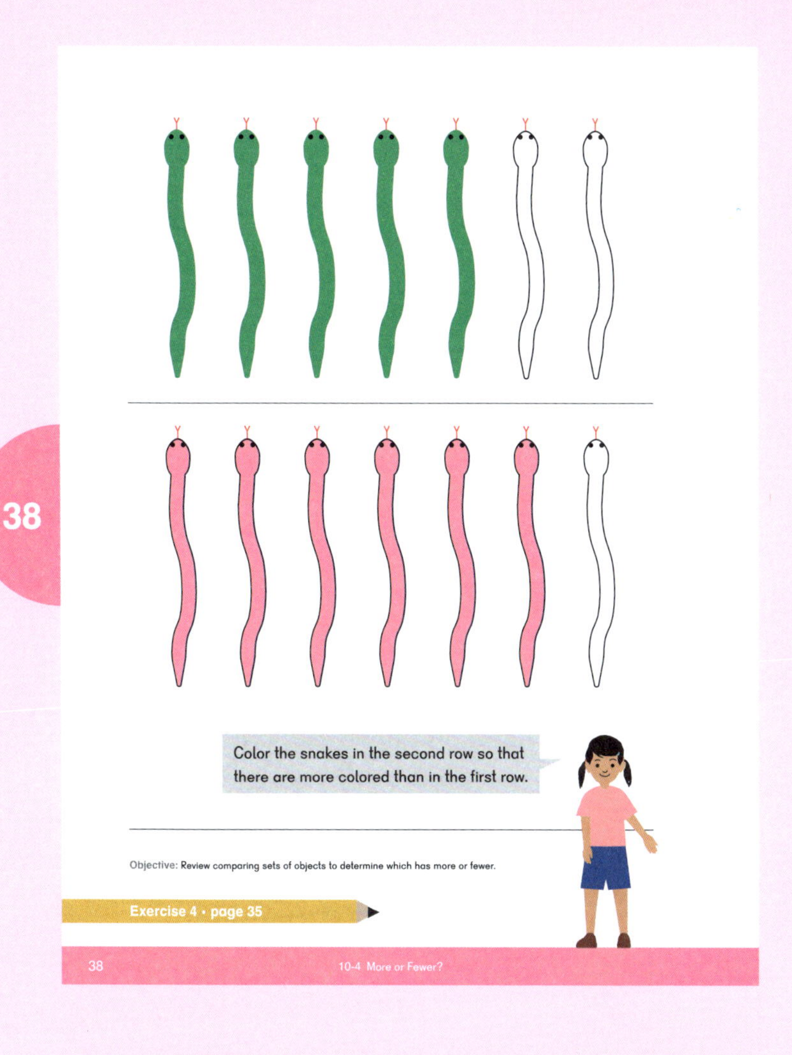

Whole Group Play

Dance and Freeze: Play music. While the music is playing, all students dance. When the music stops, students must decide whether to sit down and freeze, or to continue standing and freeze. After each round, ask if there are more students standing or sitting and how they know. Observe whether students suggest counting each group and comparing the two numbers or having each group line up and matching a sitting student to a standing student. Be sure students say that because there are more students standing (or sitting), there are fewer students sitting (or standing).

Small Group Center Play

Ten-Frame Flash: Hold up 2 Ten-frame Cards (BLM) at a time and have students point to the card that shows more frames colored in, then to the card that shows fewer frames colored in.

Dramatic Play: Have children dress up in more or fewer articles of clothing than a doll or stuffed toy you dressed.

Art: Provide play dough. Make 5 play dough snakes and display them. Have students make more, then fewer than 5 snakes.

Extend Play

Dance and Freeze with a Twist: After each round, have students say how many students are standing and sitting and say, "______ (number of students standing or sitting) is more/fewer than ______ (number of students standing or sitting)."

Exercise 4 • page 35

Objective

- Practice concepts introduced in this chapter.

Lesson Materials

- *The Very Hungry Caterpillar* by Eric Carle
- Cups and saucers for tea party
- Paper plate, construction paper, pipe cleaners to make cat face
- Optional snack: celery sticks, yogurt or cream cheese, cherry tomatoes, raisins, grapes

For the **Practice**, read the directions and speech bubbles on each page and have students complete the tasks.

Whole Group Play

Read *The Very Hungry Caterpillar* by Eric Carle. Have students discuss what the caterpillar ate, using the words "more" and "fewer." For example, the caterpillar ate more oranges than ice cream cones. Show students a "caterpillar" that you have created by filling a celery stick with yogurt or cream cheese, using cherry tomatoes or grapes instead of pom-poms for the body, and placing raisins for the eyes. Ask students if there are more grapes or raisins. Allow students to create their own "caterpillars" and enjoy.

Small Group Center Play

Tea Party: Have students set up a tea party with a different number of cups and saucers. Have them tell whether there are fewer cups or saucers.

Cat Face: Have students use paper plates, paper cutouts of triangles for ears and circle for eyes, and pipe cleaners for whiskers to make a cat face. Tell them to make their cat faces have more whiskers than eyes.

39

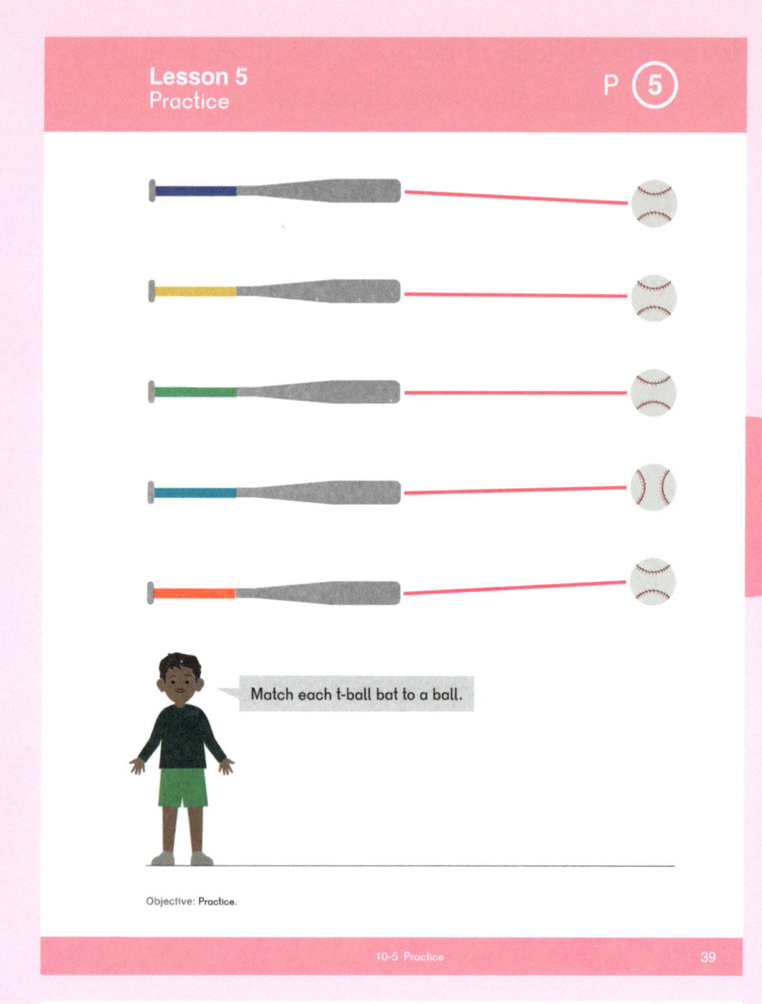

40

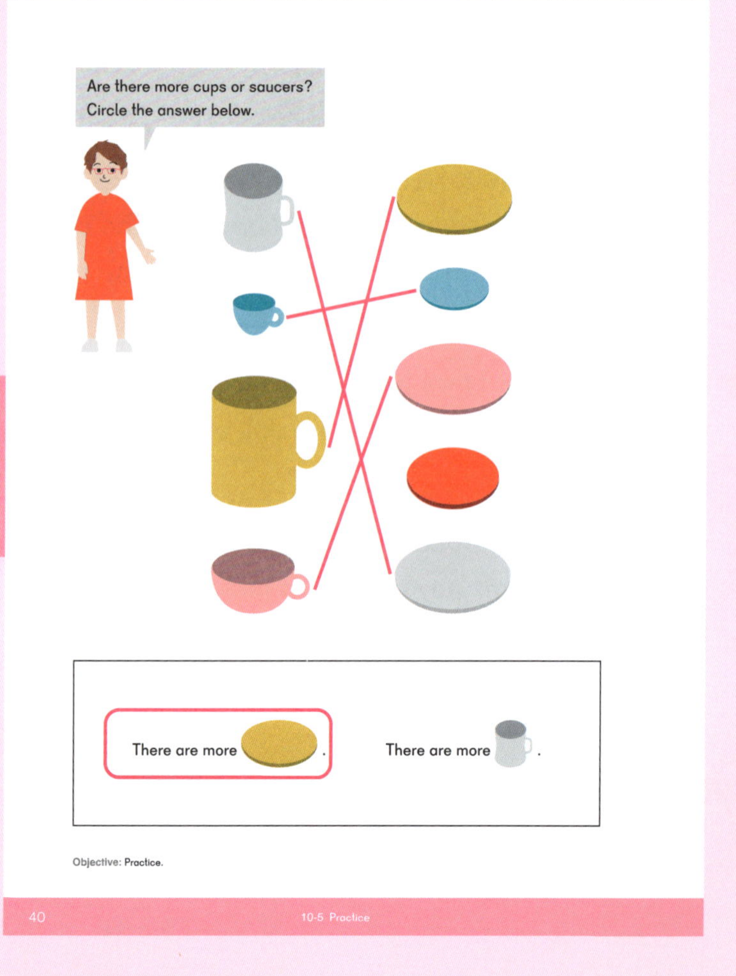

Exercise 5 • page 37

Extend Play

More or Fewer Caterpillars: Have students create two caterpillars using the materials provided. One of the caterpillars must have fewer fruit or vegetable pieces than the other. Have them identify which has more and which has fewer pieces. Enjoy!

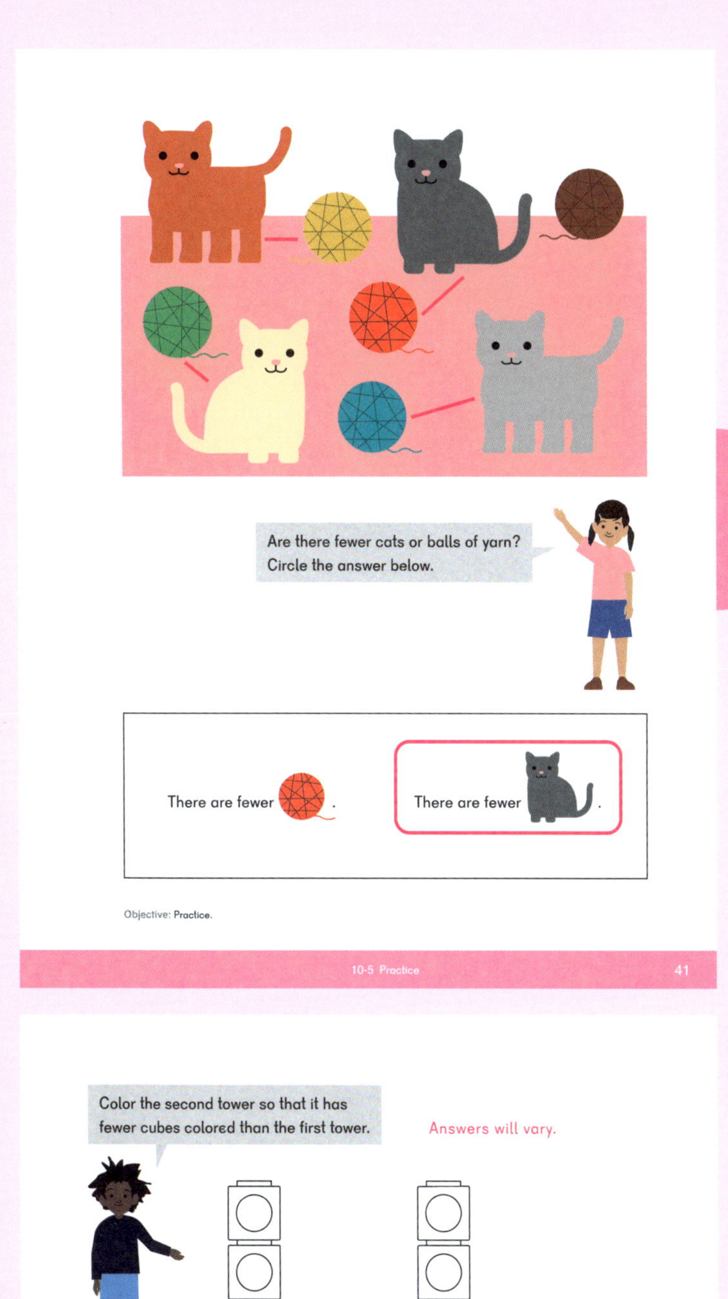

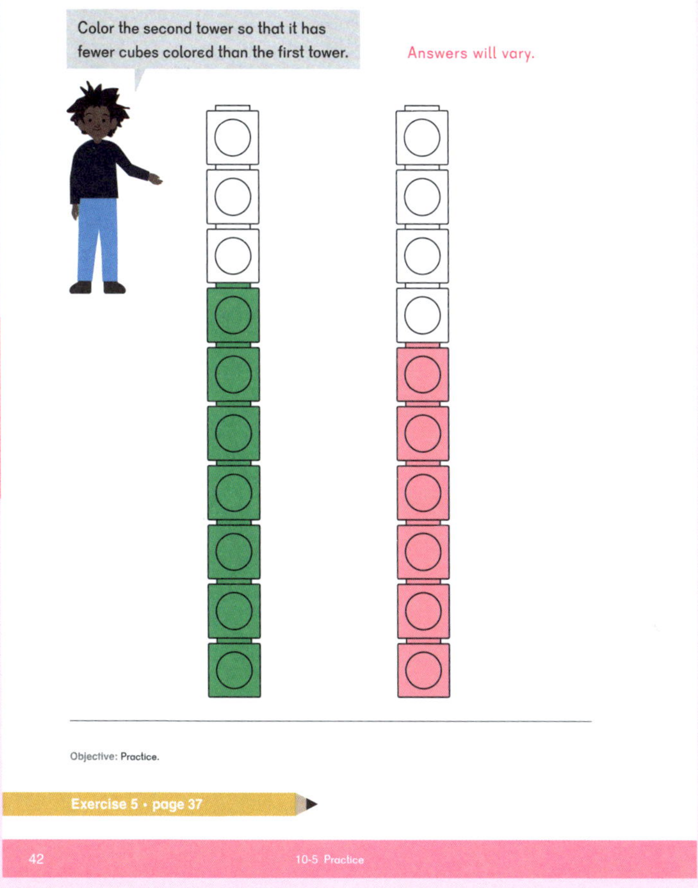

Exercise 1 • pages 27–30

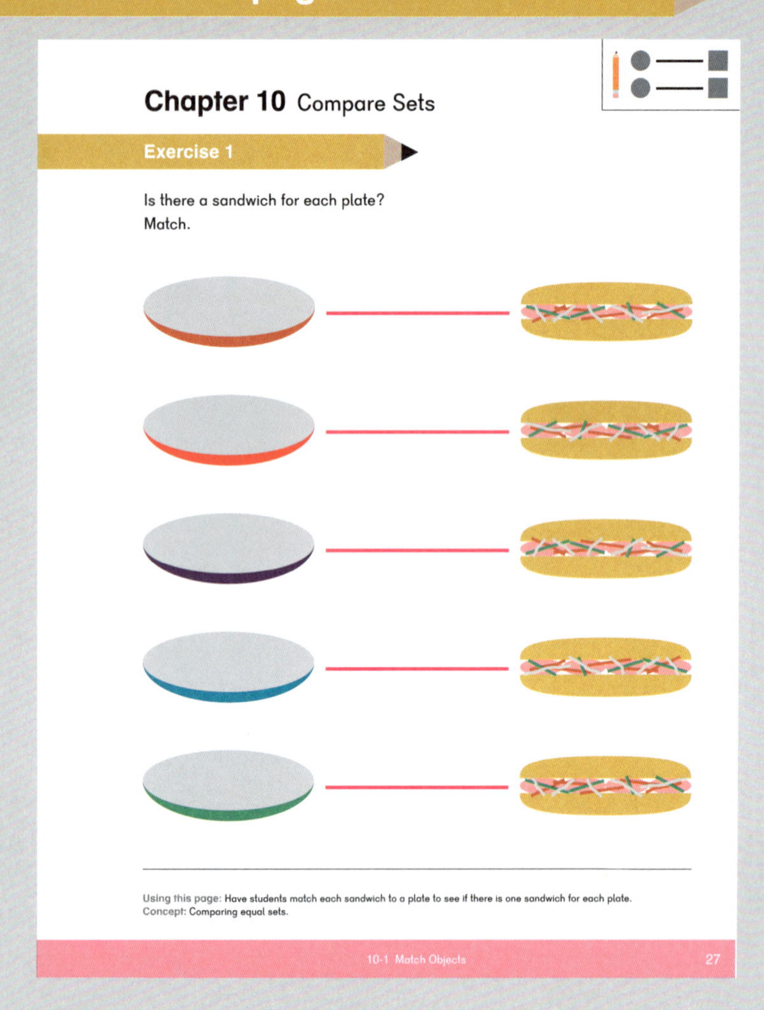

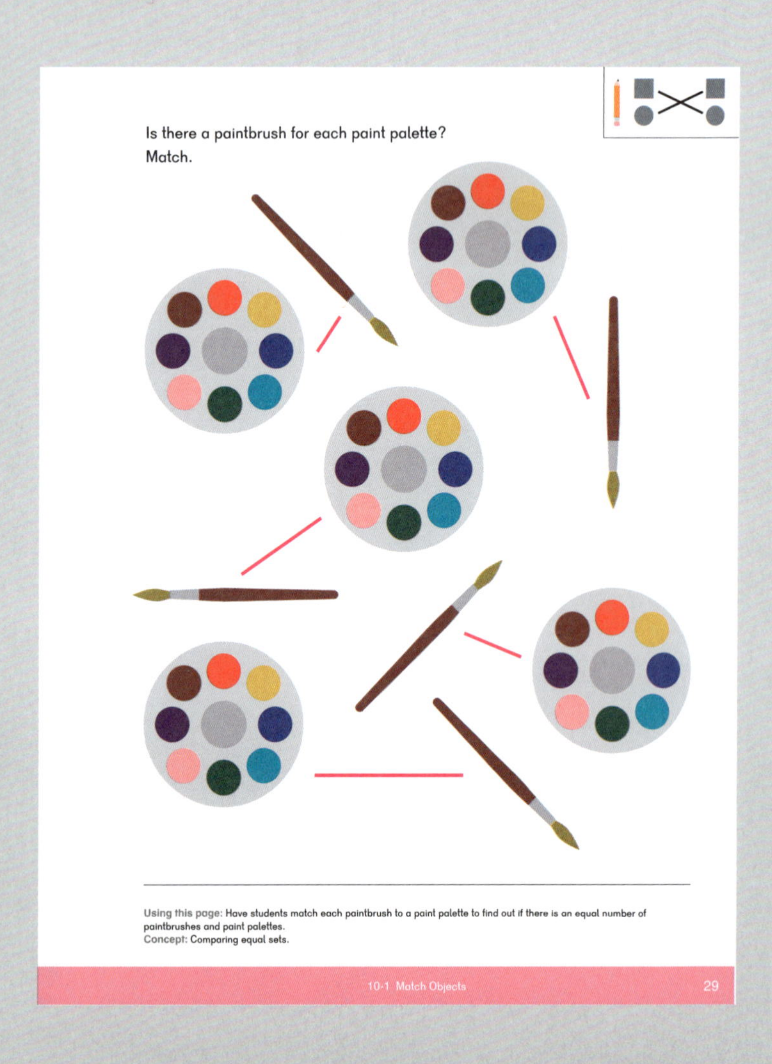

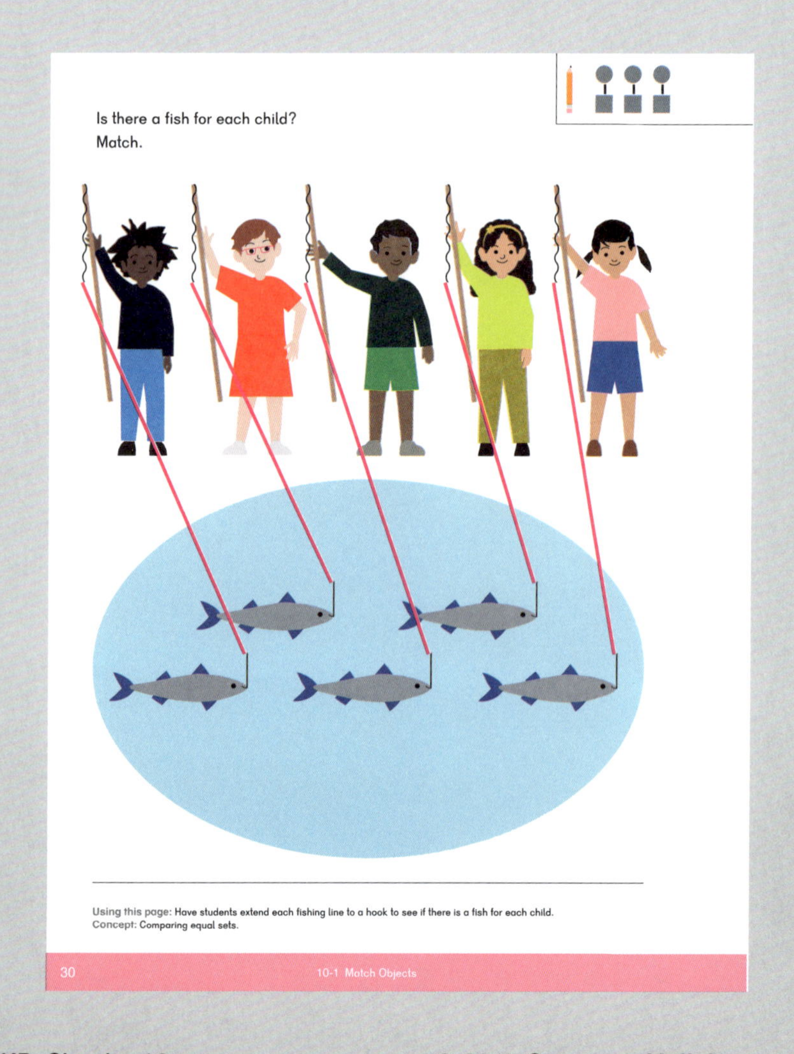

Exercise 2 • pages 31–32

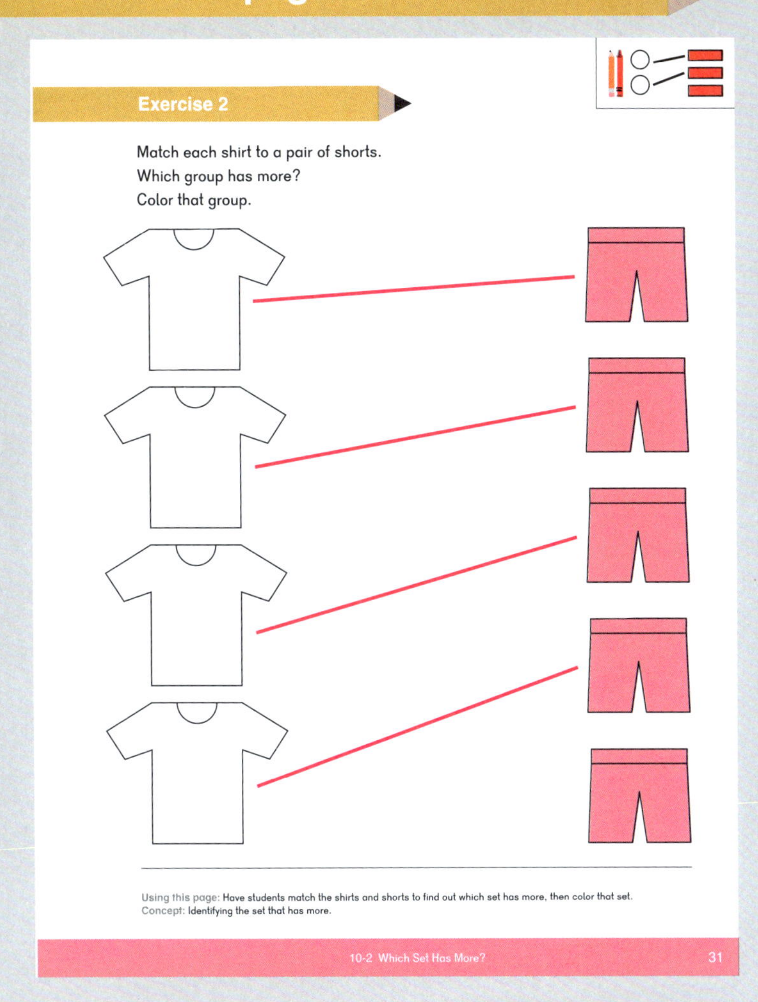

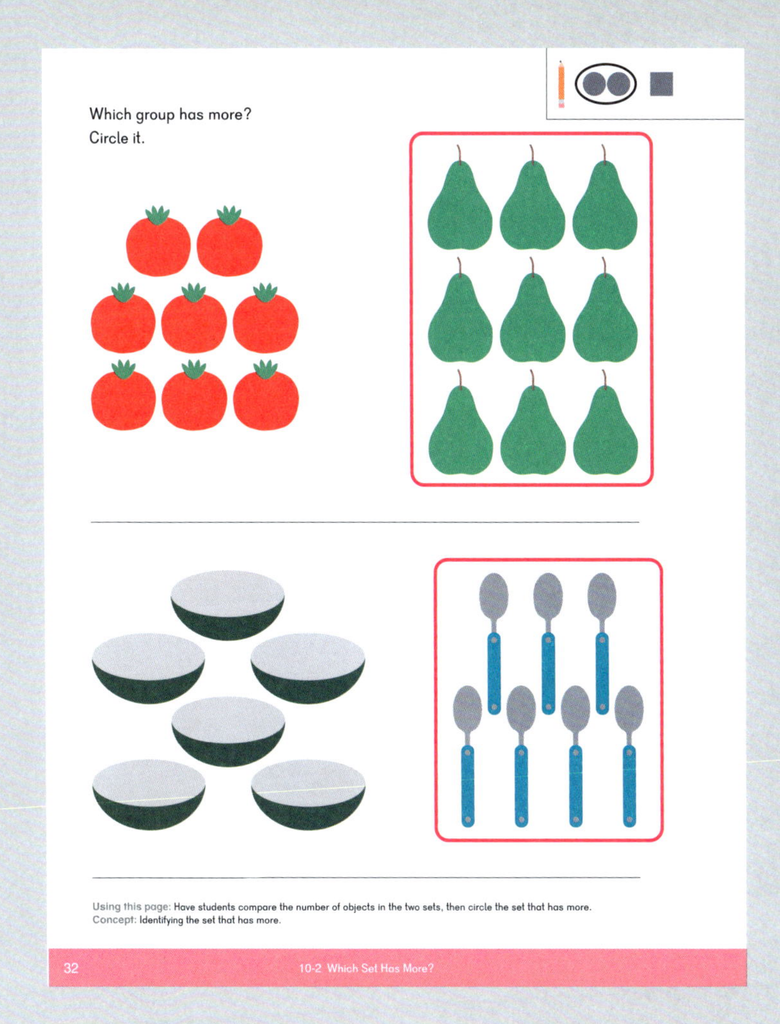

Exercise 3 • pages 33–34

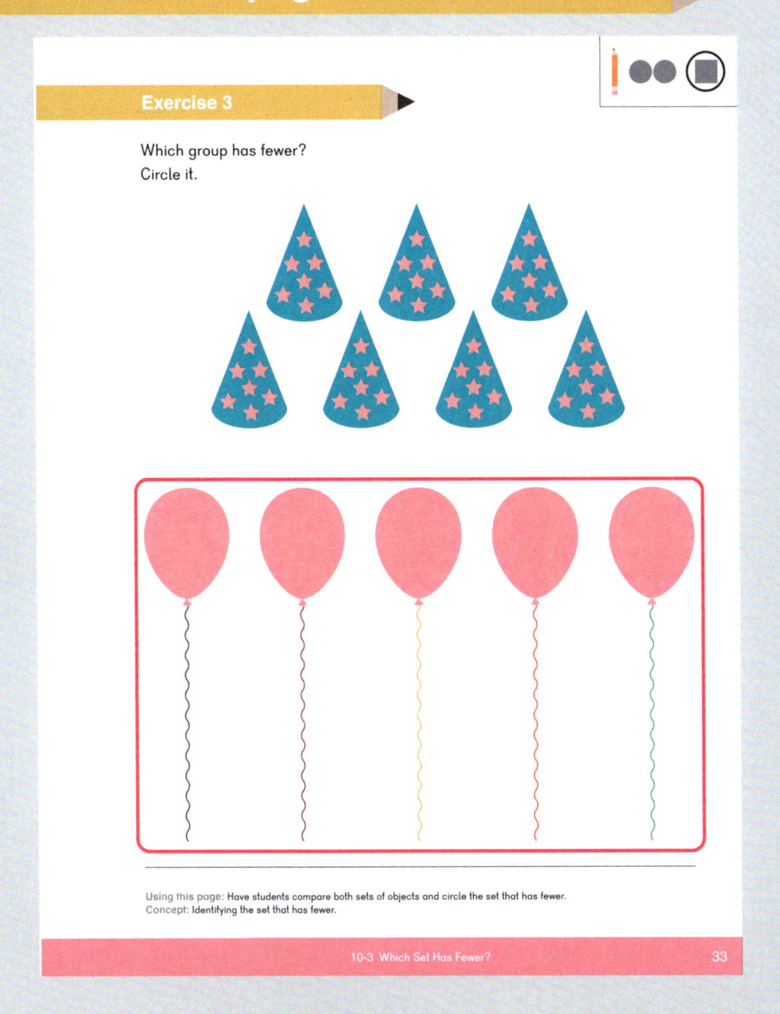

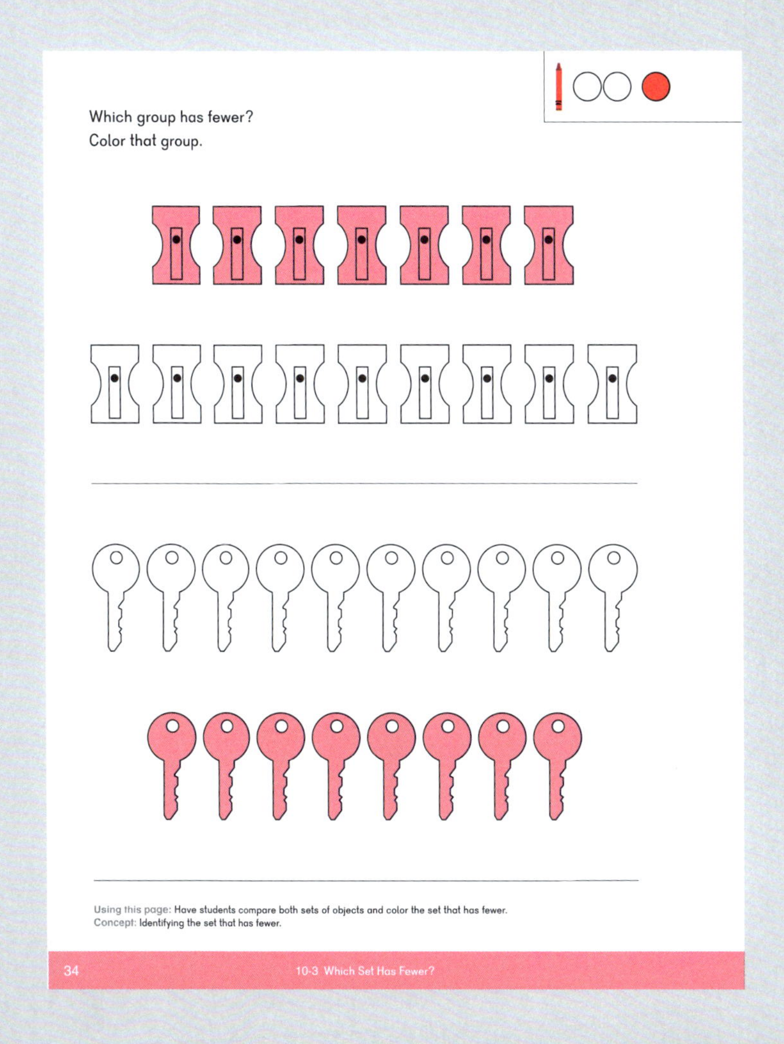

Exercise 4 • pages 35–36

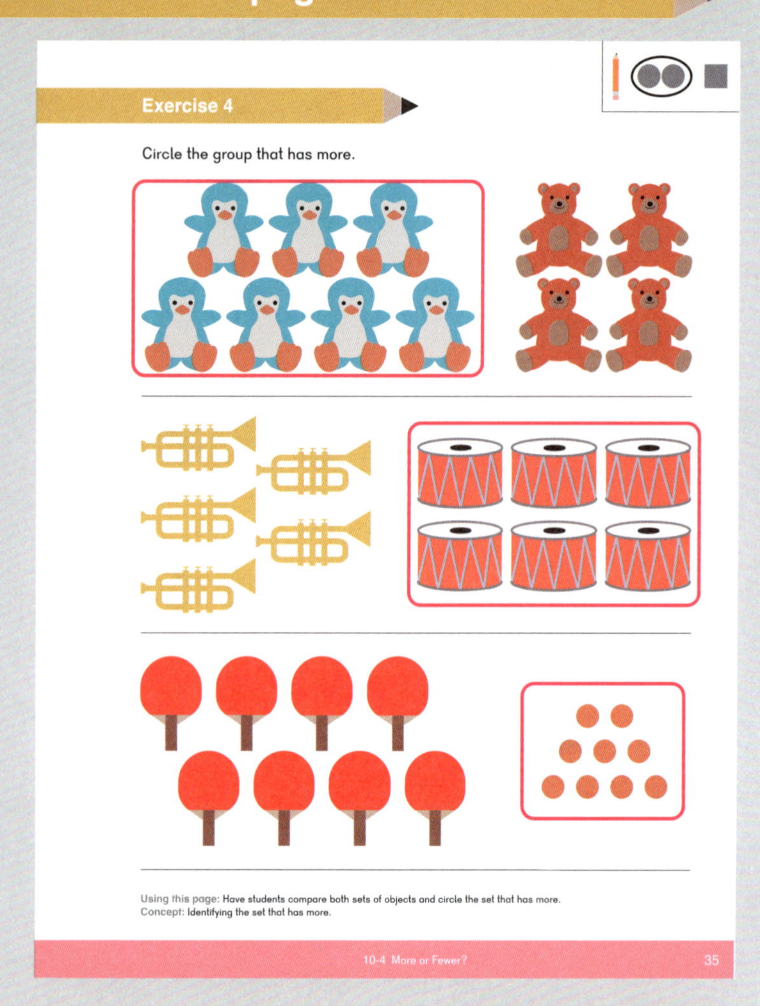

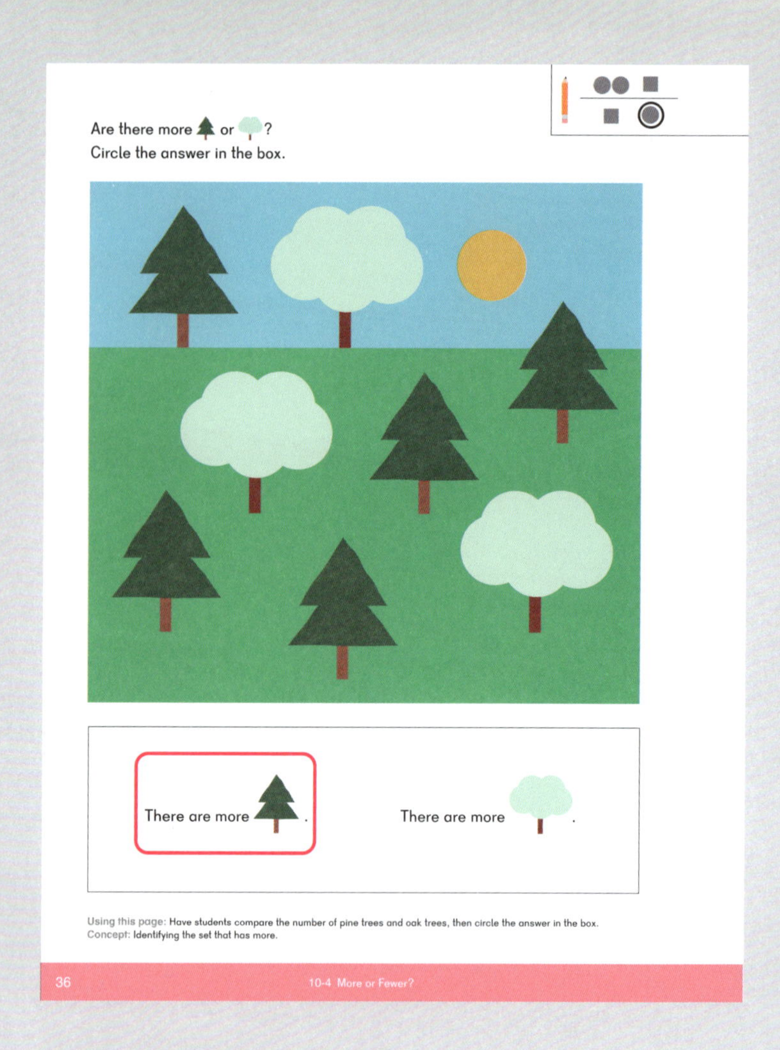

Exercise 5 • pages 37–38

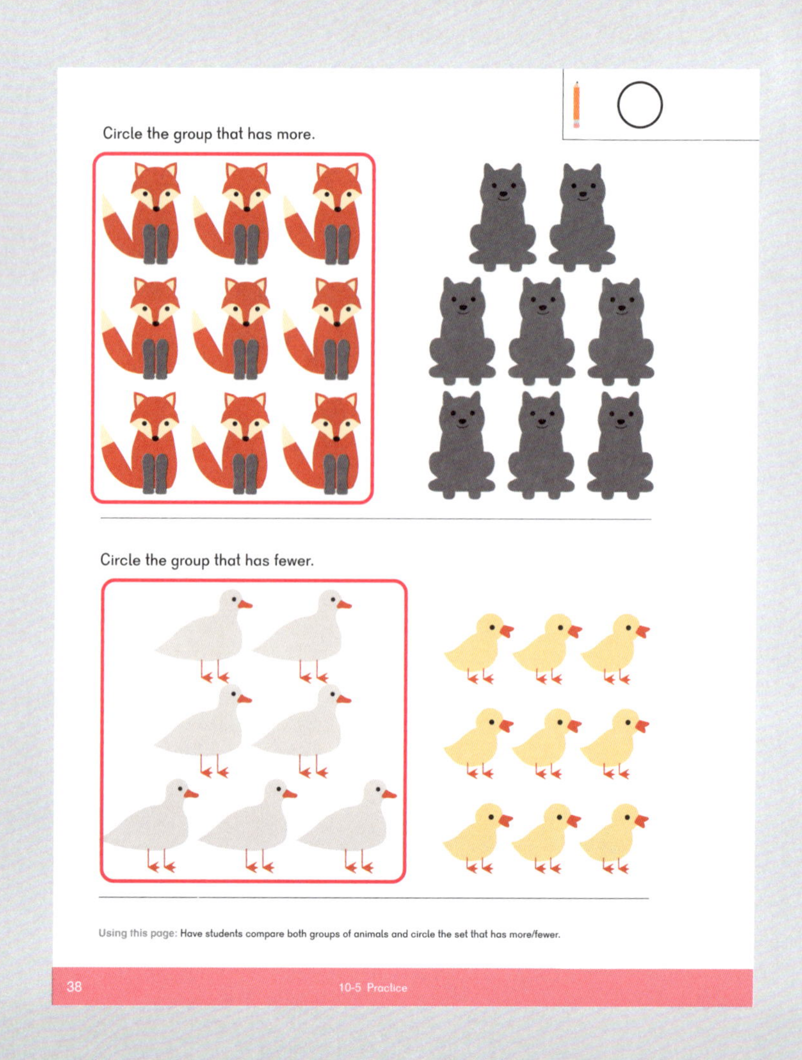

Suggested number of class periods: 6–7

Lesson	Page	Resources	Objectives
Chapter Opener	p. 67	TB: p. 43	
1 Altogether — Part 1	p. 68	TB: p. 44 WB: p. 39	Compose numbers to 5.
2 Altogether — Part 2	p. 69	TB: p. 45 WB: p. 41	Compose numbers to 5.
3 Show Me	p. 72	TB: p. 48 WB: p. 43	Identify sets of objects. Record compositions of numbers to 5 using five-frame cards.
4 What's the Other Part? — Part 1	p. 75	TB: p. 51 WB: p. 45	Find the missing part of numbers within 5 when 1 part is given.
5 What's the Other Part? — Part 2	p. 77	TB: p. 53 WB: p. 49	Find the missing part of numbers within 5 when 1 part is given.
6 Practice	p. 79	TB: p. 55 WB: p. 53	Practice concepts introduced in this chapter.
Workbook Solutions	p. 82		

Chapter Vocabulary

- Put together
- In all
- Altogether
- Missing part
- Take away
- Set
- Group

This chapter is an informal introduction to putting together and taking apart. Having already learned to count and compare quantities, students will begin to join two sets to find how many in all, and to take apart a set to find how many are left.

Students will begin by putting sets of objects together to find how many in all. They will then start with a whole and one of the parts, and will find the missing part.

Key Points

Students will be making up number stories after looking at pictures in the text. Pay close attention to make sure the story they tell matches the picture. This chapter uses informal language. In **Chapter 12: Explore Addition and Subtraction**, students will begin to use more formal terms. The words "compose" and "decompose" are for your use and will not be introduced to students. Jumping into formal math vocabulary too early can limit students' natural curiosity and critical thinking skills.

Number bonds show the part-whole relationship in addition and subtraction. An example for a number bond for a whole of 5 is shown below.

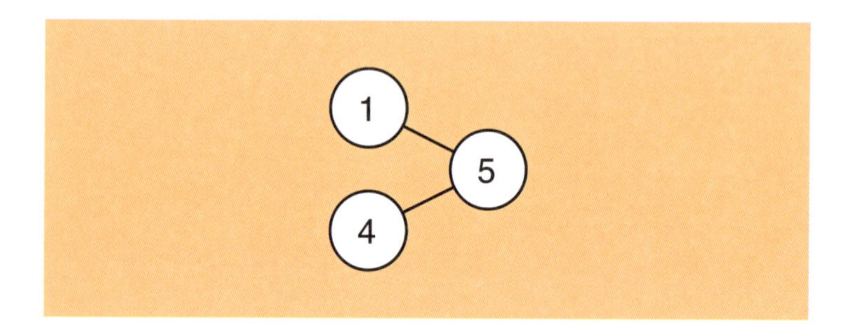

Because students are not expected to write numerals in **Dimensions Math® PK**, they will not be writing number bonds. Instead, they will say what numbers to put into the parts or the whole of bonds shown to them.

Some students may find it challenging to understand that a "part" of a "whole" can be 0. In **Lesson 4: What's the Other Part? — Part 1**, students play with dominoes and/or domino cards. Dominoes often show 0 as a part. For students who struggle, emphasize the concept using different manipulatives, such as counters and five-divot egg cartons.

Beginning with **Lesson 2**, include the following finger play rhyme:

Ten Fingers

I have ten fingers (hold up both hands, fingers spread)
And they all belong to me, (point to self)
I can make them do things —
Would you like to see?
I can shut them up tight (make fists)
I can open them wide (open hands)
I can put them together (place palms together)
I can make them all hide (put hands behind back)
I can make them jump high (hands over head)
I can make them jump low (touch floor)
I can fold them up quietly (fold hands in lap)
And hold them just so.

Explore for **Lesson 5** requires puzzle cards to be copied and cut out before the lesson.

In **Lesson 5**, encourage students to visualize the number stories you tell them. Building visualization skills at an early age will help students as they progress in their problem solving skills in upper grades. For example, use terms like "imagine" and "make a picture in your mind."

Use the terms "in all" and "altogether" interchangeably.

Materials

- Beads
- Bean bags
- Bear counters
- Counters
- Crayons
- Dice
- Dominoes
- Felt board
- Felt counters
- Hula hoops
- Linking cubes
- Pipe cleaners
- Plastic bags
- Removable dot stickers in at least 2 colors
- Small containers
- Small cups
- Small index cards
- Small items to sort
- Stickers
- Water balloons

Note: Materials for Activities will be listed in detail in each lesson.

Blackline Masters

- Blank Five-frames
- Domino Cards
- Ladybug Coloring Sheets
- Ladybug Playing Cards
- Number Cards
- Number Cards — Large
- Parts and Whole Mats

Storybooks

- *The Very Hungry Caterpillar* by Eric Carle
- *More, Fewer, Less* by Tana Hoban

Optional Snacks

- Fish crackers
- Teddy bear crackers
- Banana slices
- Dominoes (graham crackers with cream cheese, with each half separated by a thin slice of cheese, and blueberries or other small fruit pieces for domino dots)
- Finger sandwiches
- Popcorn served in plastic gloves

Letters Home

- Chapter 11 Letter

Five Little Monkeys

Five little monkeys jumping on the bed.
One fell off and bumped his head.
Mama called the doctor. The doctor said,
"No more monkeys jumping on the bed!"

Four little monkeys jumping on the bed.
One fell off and bumped his head.
Mama called the doctor. The doctor said,
"No more monkeys jumping on the bed!"

Three little monkeys jumping on the bed.
One fell off and bumped his head.
Mama called the doctor. The doctor said,
"No more monkeys jumping on the bed!"

Two little monkeys jumping on the bed.
One fell off and bumped his head.
Mama called the doctor. The doctor said,
"No more monkeys jumping on the bed!"

One little monkey jumping on the bed.
He fell off and bumped his head.
Mama called the doctor. The doctor said,
"No more monkeys jumping on the bed!"

Lesson Materials

- Bear or other animal counters, 5 per student
- 2 different small containers per student, such as 2 bowls of different colors

Explore

Hand out materials. Explain to students that they are to pretend that the counters are children, and that one container is a sandbox and the other is a swimming pool.

Tell this story: "There were 5 children playing altogether. Some of them were playing in a sandbox and some of them were playing in a swimming pool. How many children did you put in the sandbox? How about in the pool? Did anyone have different numbers in the sandbox and the pool? Are there still 5? How is that?"

Draw number bonds with the parts labeled "sandbox" and "swimming pool" to record student solutions. As you draw each number bond, retell the story. For example, "You have 3 children in the sandbox and 2 children in the swimming pool. Altogether you have 5 children."

Learn

Have students look at page 43 and discuss the picture. Tell them a put together story about the dogs, such as, "There are 3 gray dogs and 2 brown dogs. There are 5 dogs altogether." Have students make up put together stories about the flowers.

Extend Learn

Have students make up an addition story about the leaves on the flowers. (This is a whole of 8.)

Objective

- Compose numbers to 5.

Lesson Materials

- Up to 5 small items, such as crayons or toy cars, in bags, 1 bag per pair of students
- Removable dot stickers in at least 2 colors
- Linking cubes
- 2 hula hoops of different colors
- Bean bags
- Optional snack: 5 teddy bear crackers

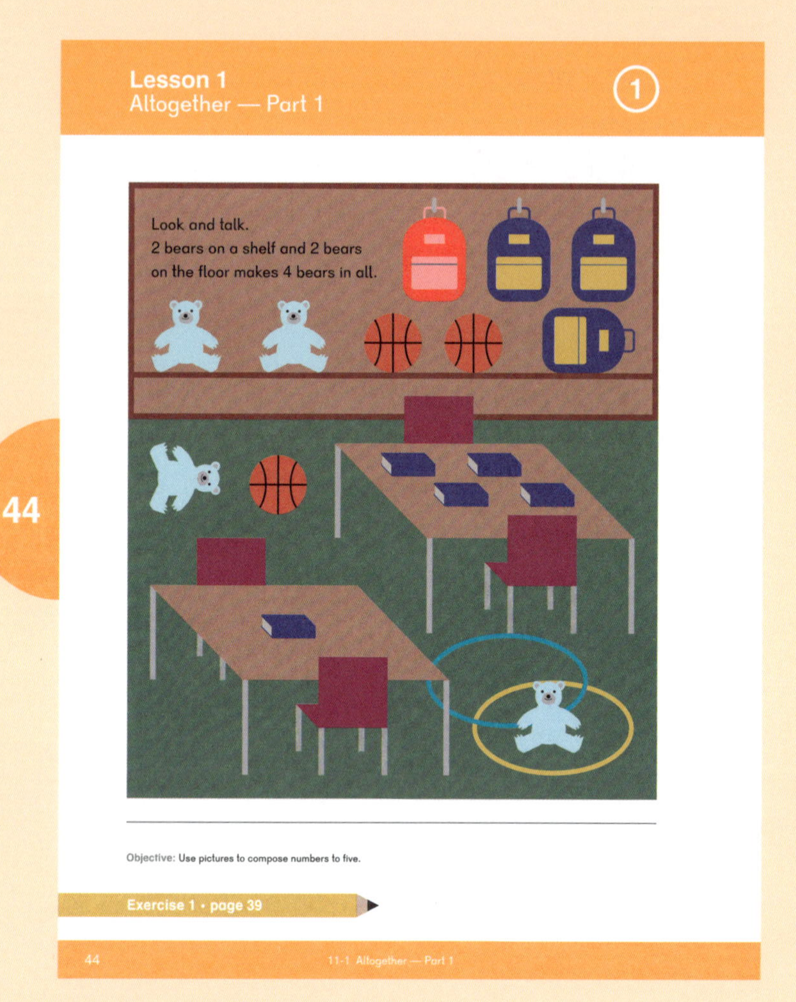

Explore

Give each pair of students a bag of objects to sort into two groups. Have students share the parts they made. Then tell them to put their parts together and tell a story about them. For example, "Two red crayons and 1 blue crayon makes 3 crayons in all."

Learn

Have students look at page 44. Hand out at least two colors of dot stickers to each student. Ask, "How many bears are on the shelf? Use one color of dot sticker to cover those teddy bears." Then say, "How many bears are on the floor? Use another color of dot sticker to cover those bears. Count all your stickers together. How many teddy bears are there in all?" Remove stickers from the bears, and repeat the series of questions for other objects on the page.

Whole Group Play

Hula Hoop Parts: Put hula hoops on the floor. Ask two students to each stand in one hula hoop. Ask questions such as, "How many students are in the pink hula hoop? How many students are in the purple hula hoop? If we put them together, how many students are there in all?" Lead them in saying, "1 student and 1 student make 2 students in all." Repeat with other students making a whole up to 5.

Small Group Center Play

Dramatic Play: Set out objects that a teacher would use, up to 5 of each, and have students play school. The "teacher" should have the "students" make up put together stories about the objects.

Outdoor Play: Have students toss bean bags into 2 hula hoops.

Exercise 1 • page 39

Extend Learn

Mystery Tower: Use linking cubes to create towers of 2, 3, 4, or 5 cubes. Give clues and have students tell how many cubes are in the tower. For example, "There are 2 brown cubes and 1 orange cube." Provide cubes for students who need them. Add positional words to your clues. Have students play in pairs.

Objective

- Compose numbers to 5.

Lesson Materials

- Number Cards — Large (BLM) 1 to 5
- Beads of 2 colors
- Felt board and counters
- Pipe cleaners
- Ladybug Coloring Sheets (BLM)
- Red and black crayons
- Ladybug Playing Cards (BLM)
- Number Cards (BLM) 2 to 5
- Optional snack: finger sandwiches

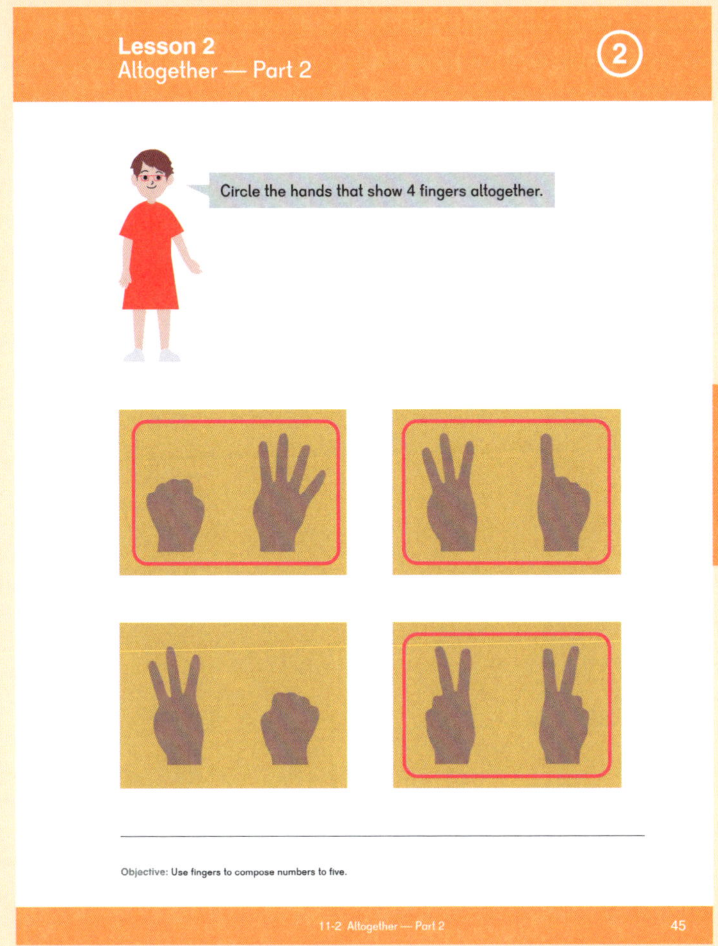

Explore

Have students work in pairs. Student One shows 1, 2, or 3 fingers, then Student Two shows 1 or 2 fingers. They then “put them together” by saying how many fingers are showing in all, for example, “1 finger and 2 fingers make 3 fingers altogether.” They repeat with different numbers of fingers, and take turns showing fingers first. Remind them that the second player can only show 1 or 2 fingers.

Learn

Hold up a Number Card (BLM) 2 to 5, and have students use fingers to show the number. Repeat the questions and statements from **Explore**.

Read Emma’s speech bubble to students and have them complete the task. After circling the correct hands, ask students to explain their thinking. Encourage them to say, for example, “3 fingers and 1 finger make 4 fingers altogether.” Be sure they recognize that 4 and 0 make 4.

Read Dion's speech bubble to students and have them complete the task. After circling the correct hands, ask students to explain their thinking. Encourage them to say, for example, "3 fingers and 2 fingers make 5 fingers in all." Be sure they recognize that 0 and 5 make 5.

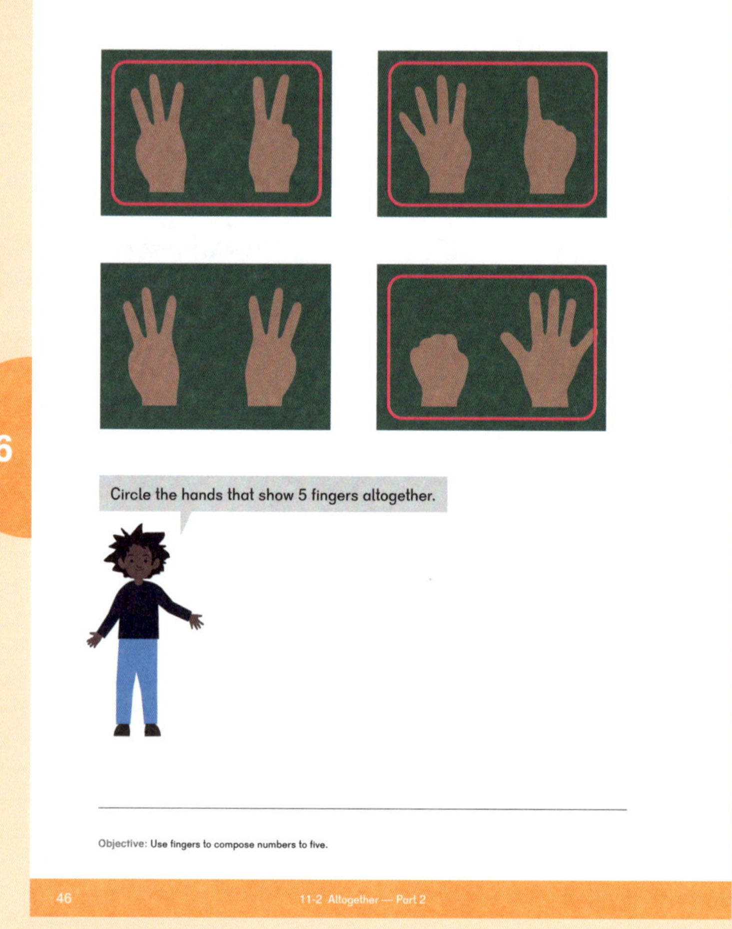

Have students talk about the dots on the ladybugs. For each ladybug, have them determine the two parts and the whole. For example, "4 dots and 1 dot make 5 dots." Read Alex's speech bubble to them and have them complete the task.

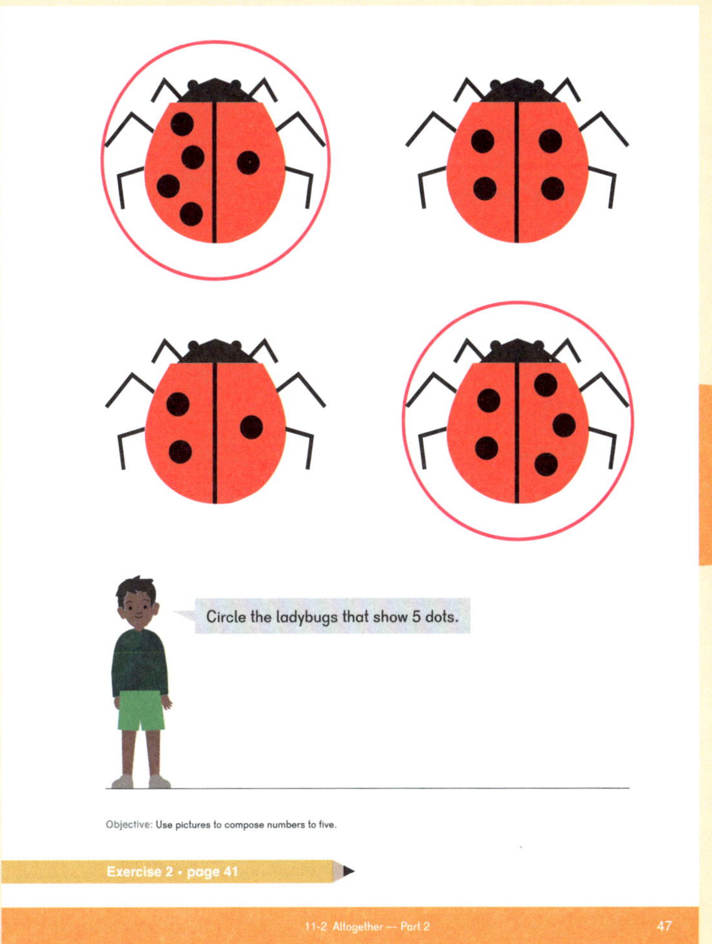

Whole Group Play

Sing "Five Little Monkeys Jumping on the Bed" (VR) with students. Model the story on a felt board and tell students to pretend that their fingers are monkeys. Have them use their fingers to show 5 on one hand at first and make a fist with the other hand. Ask, "How many monkeys are on the bed? How many are on the floor? Put them together. How many monkeys in all?" As each monkey falls off, have students show 1 fewer finger on 1 hand and 1 more finger on the other hand. After each verse, have them say, for example, "4 monkeys and 1 monkey make 5 monkeys altogether."

Small Group Center Play

Altogether 5 Bracelets: Have students create bracelets using a pipe cleaner and 5 beads of 2 different colors.

Ladybug Colors: Provide Ladybug Coloring Sheets (BLM), red crayons, and black crayons. Have students draw dots (or use bingo daubers) to show a given number of dots on the ladybugs.

Ladybug Match: Set out Ladybug Playing Cards (BLM) and Number Cards (BLM) 2 to 5. Students' task is to match the total number of dots shown on a Ladybug Playing Card (BLM) with a Number Card (BLM).

Exercise 2 • page 41

Extend Play

Ladybug Match (With a Twist): Have students match the Number Cards (BLM) showing the dots on each side of the ladybug and the numeral showing the total number of the dots to each card.

Objectives

- Identify sets of objects.
- Record compositions of numbers to 5 using five-frame cards.

Lesson Materials

- Up to 5 small objects in bags, 1 bag per pair of students (as in previous lesson)
- Blank Five-frames (BLM), 1 per pair of students
- Dominoes or Domino Cards (BLM) with wholes from 1 to 5, 1 set per pair of students
- Die with modified sides: cover the 6
- Stickers and small index cards
- 2 crayons of different colors per pair of students
- Linking cubes, 5 each of 2 different colors in plastic bags, 1 bag per pair of students
- Number Cards (BLM) 1 to 5
- Optional snack: dominoes – graham crackers with cream cheese, with each half separated by a thin slice of cheese, and blueberries or other small fruit pieces for domino dots

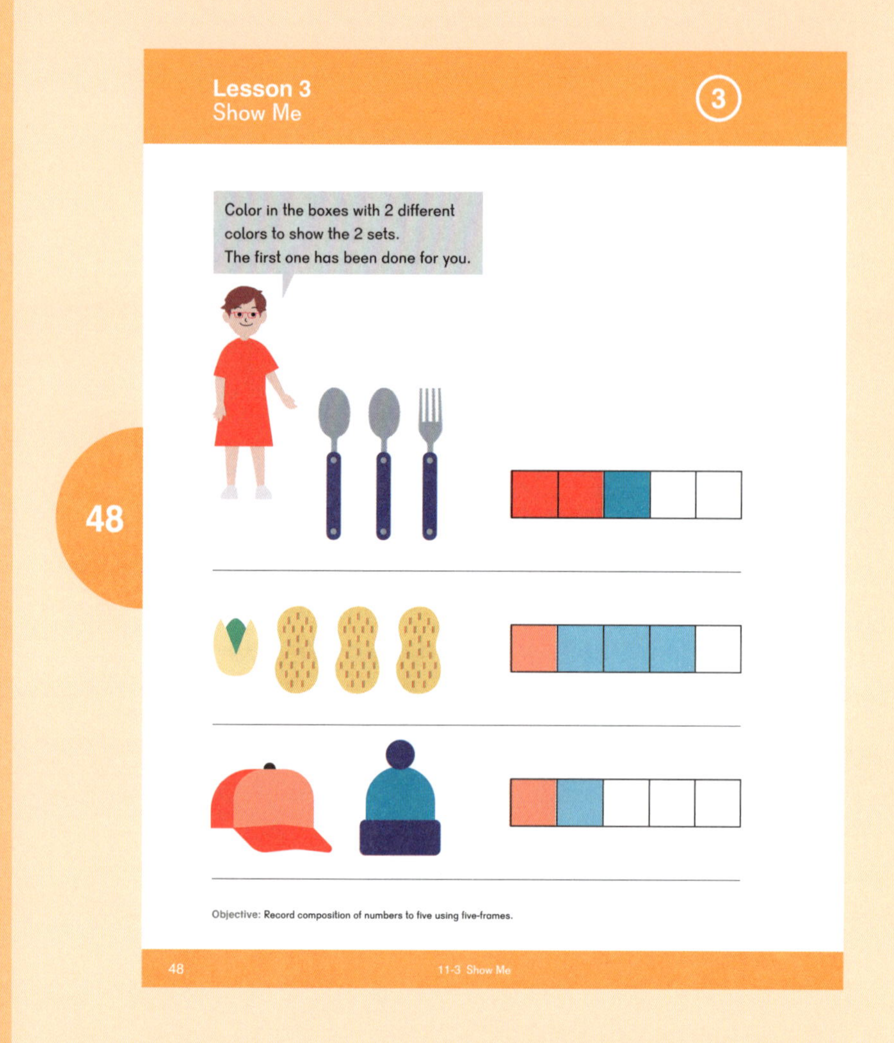

Explore

Give each pair of students a bag of small objects, a Blank Five-frame Card (BLM), and 2 crayons. Have students sort their objects into 2 groups and share the reason for the sort. Introduce the words "part" and "whole" by showing 2 groups of objects that you sorted and telling them that there are ___ objects in this part, ___ objects in this other part, and when we put them together, there are ___ objects. The whole group has ___ objects. Move the objects in the two parts together into a pile while saying, "Part and part make whole."

Learn

Introduce the word "set." Tell students that "set" and "group" will be used interchangeably. During the remainder of the lesson, have students state, "Part and part make whole," for each problem or story. For example, on page 48, for the first row, "2 and 1 make 3."

Model for students how to color in a Blank Five-frame Card (BLM) to record compositions. For example, for 2 big paper clips and 1 small paper clip, color 2 sections of the five-frame using 1 color and 1 section of the five-frame using a different color. Repeat with different sets of objects, and have students tell you how to color the Blank Five-frame Card (BLM).

Have pairs of students use crayons to color their Blank Five-frame (BLM) to represent their sets of objects. If necessary, have them put counters on their frames first, lift the counters, and then color.

Call on several students to have them share their work.

Have students look at page 48 and read Emma's speech bubble. Ask them to identify the objects in the first row (spoons and forks). Tell why two sections of the five-frame are colored 1 color and 1 section of the five-frame is colored a different color. Have them identify the objects in the second row (pistachio and peanuts) and tell how they will color the five-frame. Have them complete the task.

Have students look at page 49. Read Emma's speech bubble. Ask them what the sets are in the first row. Have them color the frames. Repeat for the second row.

Ask what is different about the objects in the third row from those in the first and second rows. (The items from the same set are not always next to each other or grouped together.) Give them linking cubes of two different colors and have them arrange and rearrange them, so that the same colored linking cubes are not always adjacent. Then have them color a Blank Five-frame (BLM) as if the items in each set are adjacent to each other. Repeat for the objects in the fourth row.

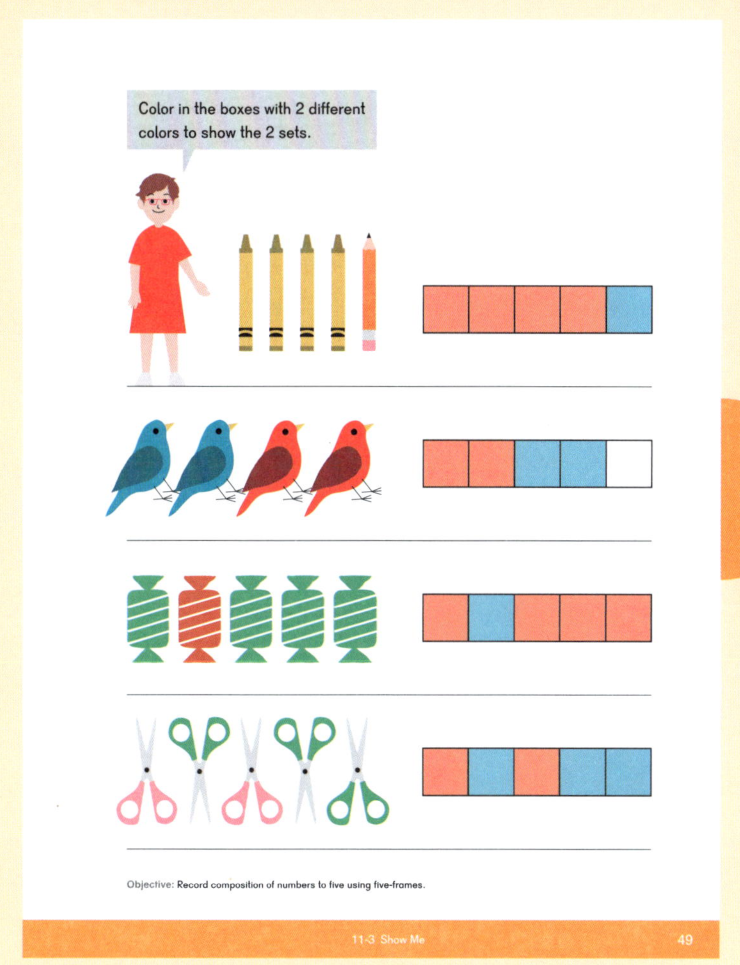

Show students a domino or Domino Card (BLM) with a whole of up to 5. Have them say, for example, "One and three make four." Repeat with other dominoes. Have students look at page 50. Read Mei's speech bubble. Have them tell how they plan to color in the Blank Five-frame Card (BLM) and which numeral they will circle. Have them complete the task.

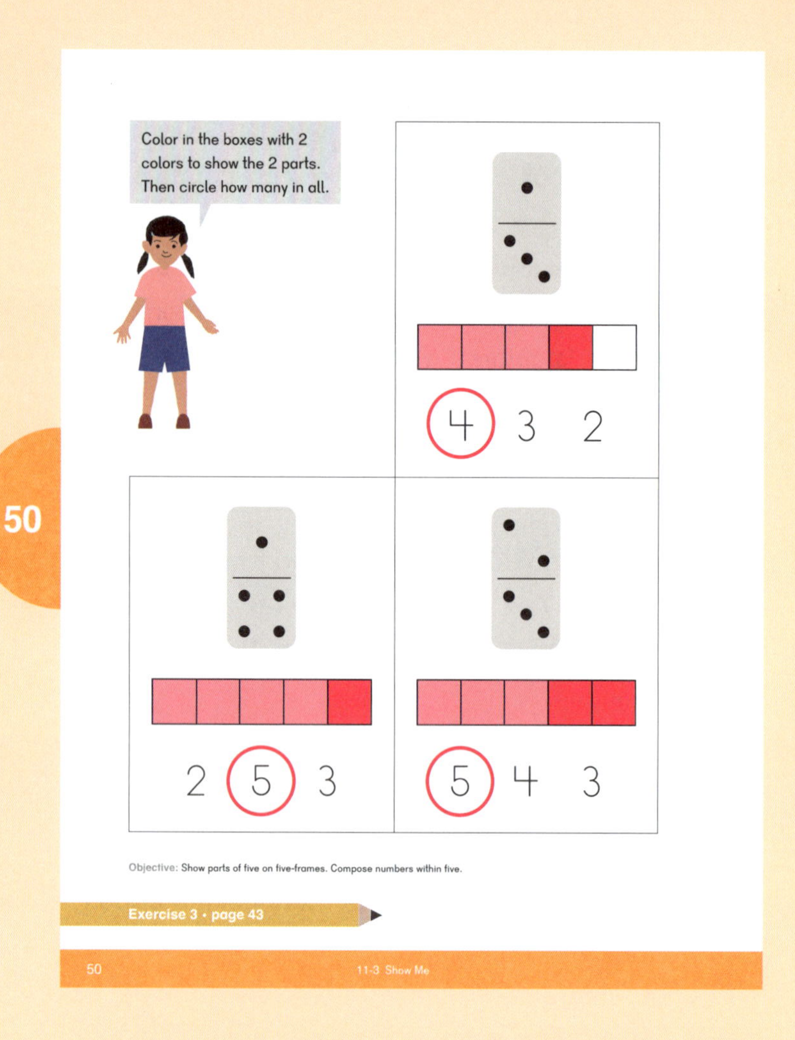

Whole Group Play

Give each pair of students a bag of linking cubes and a Blank Five-frame Card (BLM). Tell a composition (altogether) story and have students model the stories with their cubes.

Possible stories:

- There are 2 books on the floor and 1 book on the table. There are 3 books in all. "2 and 1 make 3."
- There were 3 stuffed animals in one basket and 1 stuffed animal in another basket. There were 4 stuffed animals altogether. "3 and 1 make 4."

Small Group Center Play

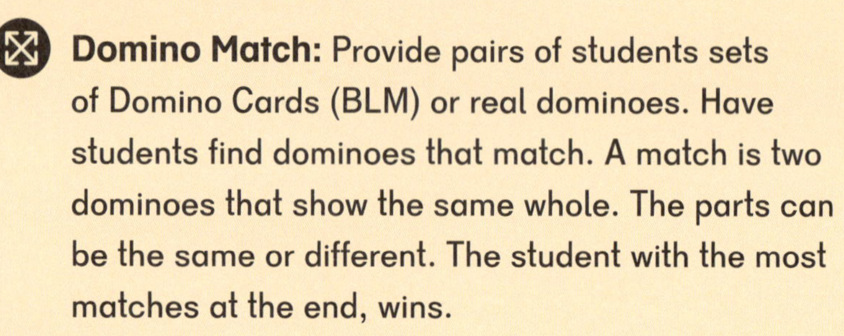

Domino Match: Provide pairs of students sets of Domino Cards (BLM) or real dominoes. Have students find dominoes that match. A match is two dominoes that show the same whole. The parts can be the same or different. The student with the most matches at the end, wins.

Domino-Dice Match: Provide pairs of students sets of Domino Cards (BLM) or real dominoes and a die with the 6 covered. Have students roll the die, then find a domino that shows the same whole as the number represented on the die.

Pretty Dominoes: Fold small index cards in half. Provide a set of dominoes or Domino Cards (BLM), stickers, and index cards. Have students create a set of dominoes to match the set provided using stickers on index cards.

Exercise 3 • page 43

Extend Play

Have students match their pretty dominoes from **Small Group Center Play** to Number Cards (BLM) showing how many in all on each domino.

Objective

- Find the missing part of numbers within 5 when 1 part is given.

Lesson Materials

- Small cups containing 5 animal counters, 1 cup per pair of students
- Blank Five-frames (BLM), 1 per student
- Optional snack: 5 fish crackers handed out in two groups per student

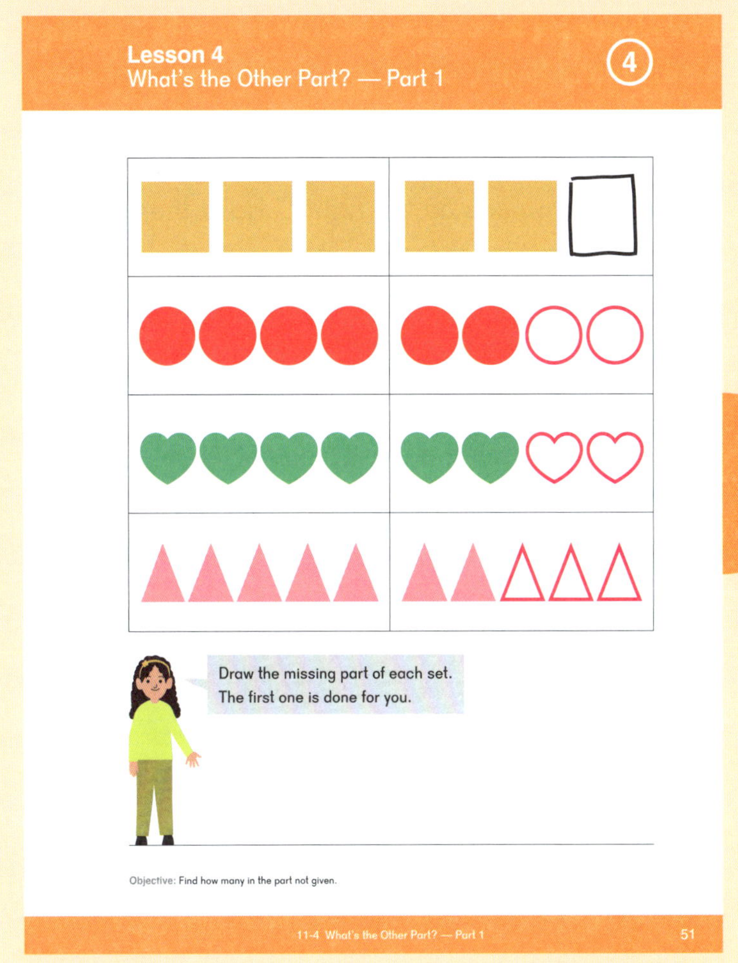

Explore

Give each pair of students a small paper cup containing 5 animal counters and a Blank Five-frame (BLM). Have students tell you how many counters they have. Then tell them that they are going to play a game called, "How Many are Hiding?" One of the students in each pair will shake up the cup and pour out the animals, while the other student closes her eyes. The first student will hide 1 or more of the counters under the cup. The other student will place the visible counters on a Blank Five-frame Card (BLM), then try to tell how many are missing. Have students take turns hiding and deducing the number missing.

Learn

Play "How Many are Hiding?" as a class.

Have students count 3 animal counters with you as you put them into a cup. Tell them that you are going to shake up the cup and pour out the animals, and that sometimes some of the animals like to hide. Have students close their eyes as you shake the cup.

Pour the animals on the floor, hiding one in your hand. Have students open their eyes. Say, "How many animals did I put in the cup? How many animals are on the floor? How many animals are hiding? How do you know? How did you decide how many were hiding?" After students answer all the questions, tell them that the whole has 3 animals, the part on the floor has 2 animals, and the missing part has 1 animal hiding. Repeat, hiding 2 animals. Then repeat with 4, then 5 animals in the whole, hiding different numbers each time, having students identify the whole, the part showing, and the missing part.

Have students look at page 51 and identify the shapes on the left side of the first row. Ask them how many squares there are. Say, "There are 3 squares on the left. Use your animal counters to show 3. How many squares do you see on the right?" Have them draw 1 square. Repeat for the other shapes. If students can draw the required shapes without using counters to model first, encourage them to do so.

Have students look at page 52. Ask them how many pears are shown in the set on the left, then how many are missing on the right. Read Mei's speech bubble and have them use their fingers to trace the circle around the 2. Ask them how many feathers are missing and which number they will circle, then have them circle the 3. Provide support as needed to complete the task.

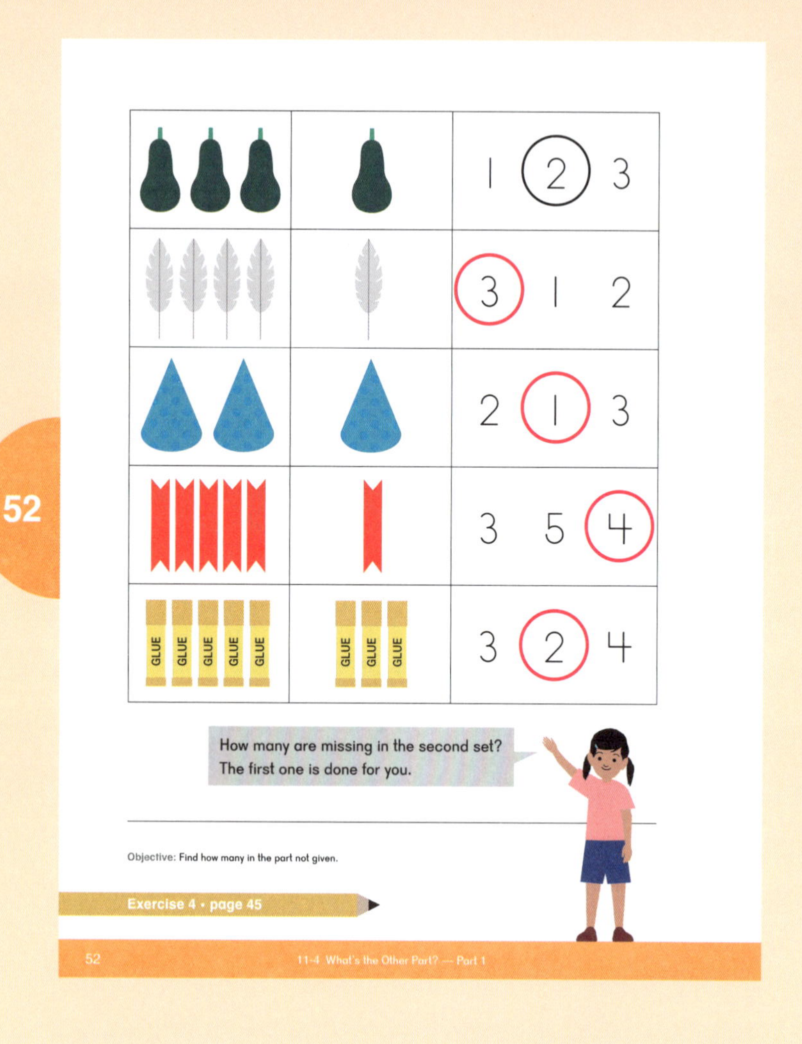

Whole Group Play

Hide and Seek: Call up groups of 3, 4 or 5. Have 1 or more of them hide and have the other students tell how many students are hiding.

Small Group Center Play

Hiding Animals: Have students pretend to be animals similar to the animal counters they used. Have some students hide and other students tell how many are hiding.

Five Little Monkeys: Play "Five Little Monkeys" (VR) and have students act it out. Each time a monkey falls off the bed, have students tell how many are still on the bed.

Animal Stories: Have students make up stories about a set of animals, up to 5, where 1 or more of them hides or gets lost. They can either draw a picture to illustrate their story or tell it aloud. If you have a recording device, they could record their stories.

Exercise 4 • page 45

Extend Learn

Finger Show: Have students play in pairs. Tell them that together, they will show 3, then 4, then 5 fingers. One student shows 1 part, then the other student must show the missing part.

Objective

- Find the missing part of numbers within 5 when 1 part is given.

Lesson Materials

- Linking cube towers of 1 to 4 cubes and of 3 to 5 cubes
- Blank Five-frame (BLM), 1 per student
- Counters of 2 different colors
- Number Cards (BLM) 0 to 5
- Optional snack: Banana slices, 5 per student

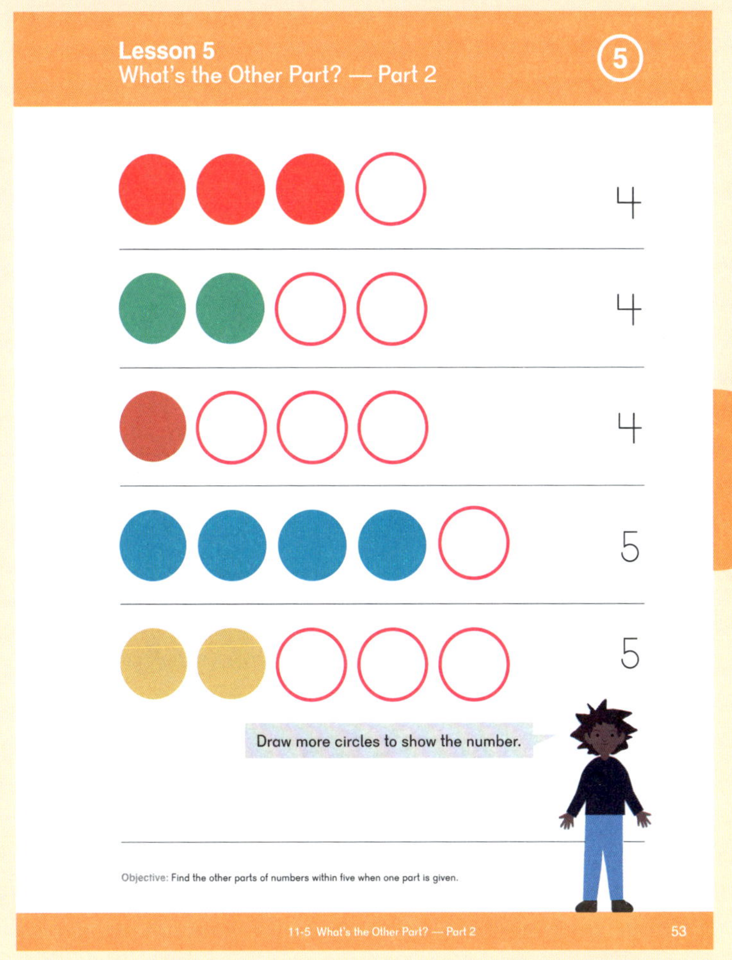

53

Explore

Give each student 1 to 4 linking cubes and have them create a tower. Then have them find a student with a tower that, when added to theirs, makes a tower of 5.

Learn

Read "Five Little Monkeys" from page 66 of this Teacher's Guide to the students. Have 5 students act out the rhyme by jumping up and down behind a curtain so that the other students can only see them when they have jumped up. When you read, "One fell off and bumped his head," have 1 student come out from behind the curtain and sit down. At this point, tell acting students to stand still and say, "There were 5 monkeys jumping on the bed. We can see the monkey who fell off the bed. How many monkeys are still on the bed?"

If necessary, suggest that students use their fingers to answer the question. Then say, "There were 5 monkeys in all. Now part of the monkeys are on the bed and part of the monkeys are on the floor. What part of 5 is on the bed? What part of 5 is on the floor? 5 is 4 and 1." Repeat until there are 0 monkeys left to jump on the bed, having students repeat, "5 is 2 and 3," "5 is 3 and 2," "5 is 4 and 1," and, "5 is 5 and 0," after each time a monkey falls off the bed.

Have students look at page 53. Ask them how many circles are shown on the top row, then what number they see next to the circles. Read Dion's speech bubble and have students complete the task.

Have students look at page 54. Read Emma's speech bubble and have them complete the task.

Whole Group Play

Snap: Give each pair of students two towers of linking cubes of 3, 4, or 5 cubes. Have them tell how many cubes are in each whole tower. Have them put one tower on a table near them to use as a visual reference of that quantity. Model for them snapping the tower into two parts and hiding one part behind your back. Have them tell how many cubes are behind your back. Give students a Blank Five-frame Card (BLM) to help them answer, if necessary. Then have pairs of students play the game together.

Small Group Center Play

I Wish I Had 5: Set out counters of 2 different colors and Number Cards (BLM) 2 to 4 facedown. Students will play in pairs. One student will say, "I wish I had 5 but I only have..." as she turns over a number card and says the number on the card. The other student may use counters or fingers, if necessary, to tell how many more are needed to get to 5. Then both students use counters to model the two parts of 5. Students take turns being the Wisher and the Teller.

Match Them: Set out Number Cards (BLM) 0 to 5 and small counters of two different colors. Students' task is to make 5 using sets of 2 cards, then model the two parts of 5 using counters.

Exercise 5 • page 49

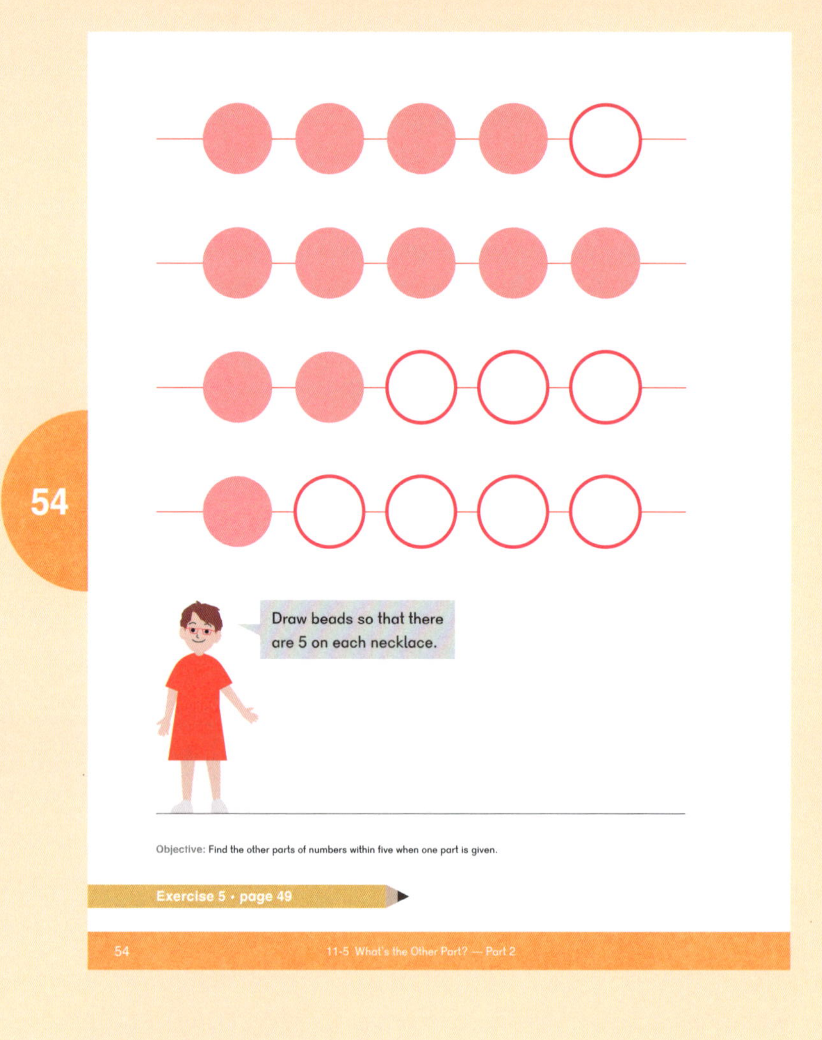

Extend Learn

Imagine This: Have students close their eyes and use mental images to answer the following questions.

- 3 children got into a minivan. The minivan could carry 5 children. How many more children could get into the minivan? (2 more children could get into the minivan.)
- Lindsay had rings on 4 of her fingers on her left hand. How many of Lindsay's fingers on her left hand did not have rings? (1 of Lindsay's fingers on her left hand did not have a ring.)

Objective

- Practice concepts introduced in this chapter.

Lesson Materials

- Linking cubes, loose and in towers up to 5
- Parts and Whole Mats (BLM)
- Domino Cards (BLM) or dominoes
- Number Cards (BLM) 0 to 5
- Small counters of 2 different colors
- Water balloons, 5 per pair of students, if weather allows
- Optional snack: popcorn served in plastic gloves

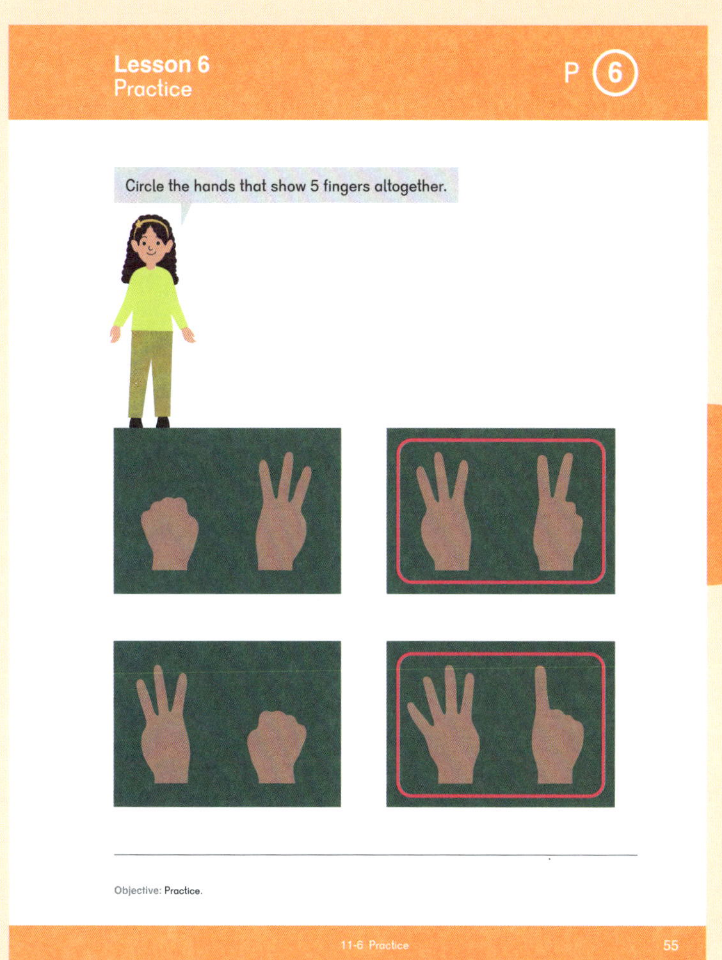

For the **Practice**, read the directions and speech bubbles on each page and have students complete the tasks.

Whole Group Play

Give each student a set of 5 linking cubes and a Parts and Whole Mat (BLM). Have students use the cubes to model these stories by placing the correct number of cubes on each part of their work mats. After modeling each story, have them say the corresponding number sentence along with you.

- 3 girls and 2 boys were eating lunch. How many children were eating lunch in all? (3 and 2 is 5.)
- 1 mama duck and 2 baby ducks were swimming. How many ducks were swimming altogether? (1 and 2 is 3.)
- 5 puppies were sleeping. 3 puppies woke up and ran to play. How many puppies were still sleeping? (5 is 3 and 2.)
- 4 children were reading books. 1 of them stopped reading to eat lunch. How many children were still reading? (4 is 1 and 3.)
- 5 superheroes were having a meeting. 2 of them flew away. How many superheroes were still at the meeting? (5 is 2 and 3.)

Hand out plastic gloves filled with popcorn. Have students work in pairs to show the correct number of fingers on 2 gloves for stories you tell similar to those above.

Small Group Center Play

Domino Match: Provide pairs of students sets of Domino Cards (BLM) or real dominoes. Have students find dominoes that match. A match is two dominoes that show the same whole. The parts can be the same or different. The student with the most matches at the end wins.

Match Them: Set out Number Cards (BLM) 0 to 5 and small counters of two different colors. Have students make 5 using sets of 2 cards, then model the two parts of 5 using counters.

Snap: Give each pair of students a tower of linking cubes. Have them tell how many cubes are in the whole tower. Model for them snapping the tower into two parts and hiding one part behind your back. Have them tell how many cubes are behind your back. Then have pairs of students play the game together.

I Wish I Had 5: Play as directed in **Lesson 5.**

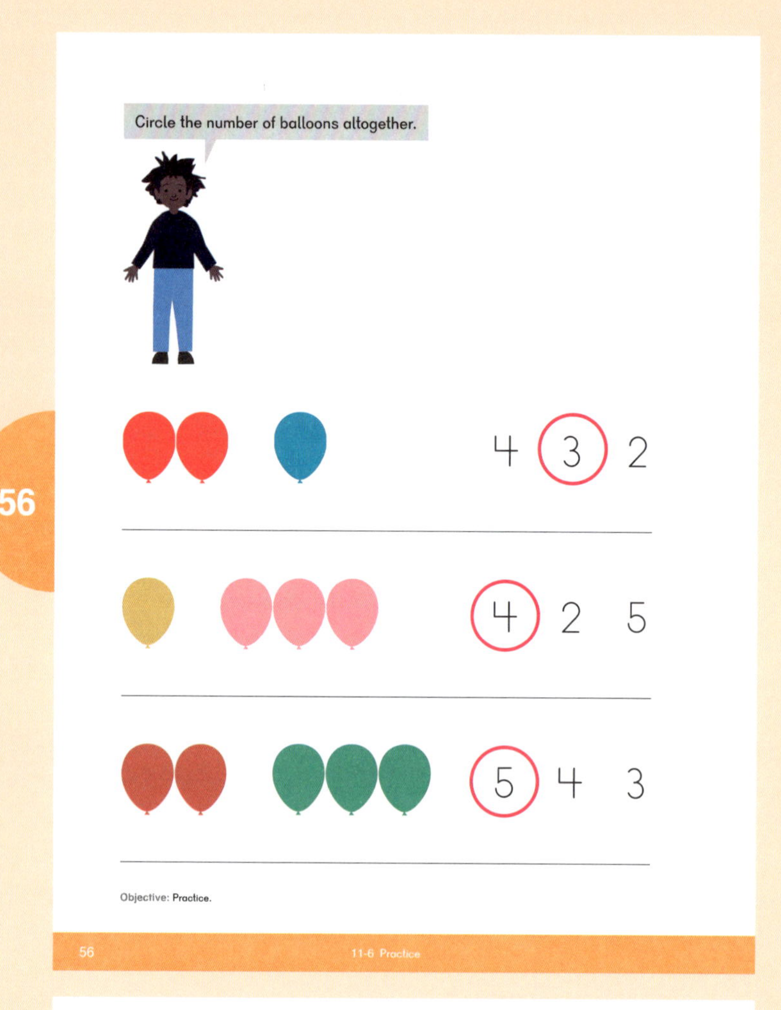

Outdoor Play

Water Balloon Toss: Give each pair of students 5 water balloons. Have them toss the balloons to each other. When a balloon bursts, ask them how many balloons they have left. Continue until all of the balloons have burst.

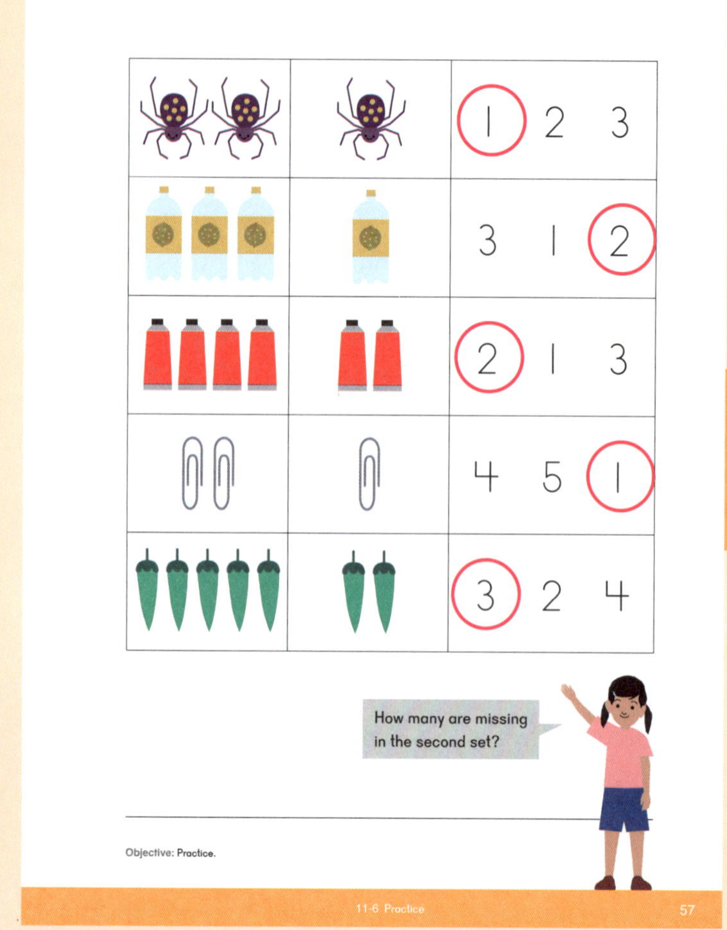

Exercise 6 • page 53

Extend Play

Tell Me a Story: Have students make up stories similar to those you told earlier. They can either draw a picture to illustrate their stories, or tell them aloud. If you have a recording device, they could record their stories.

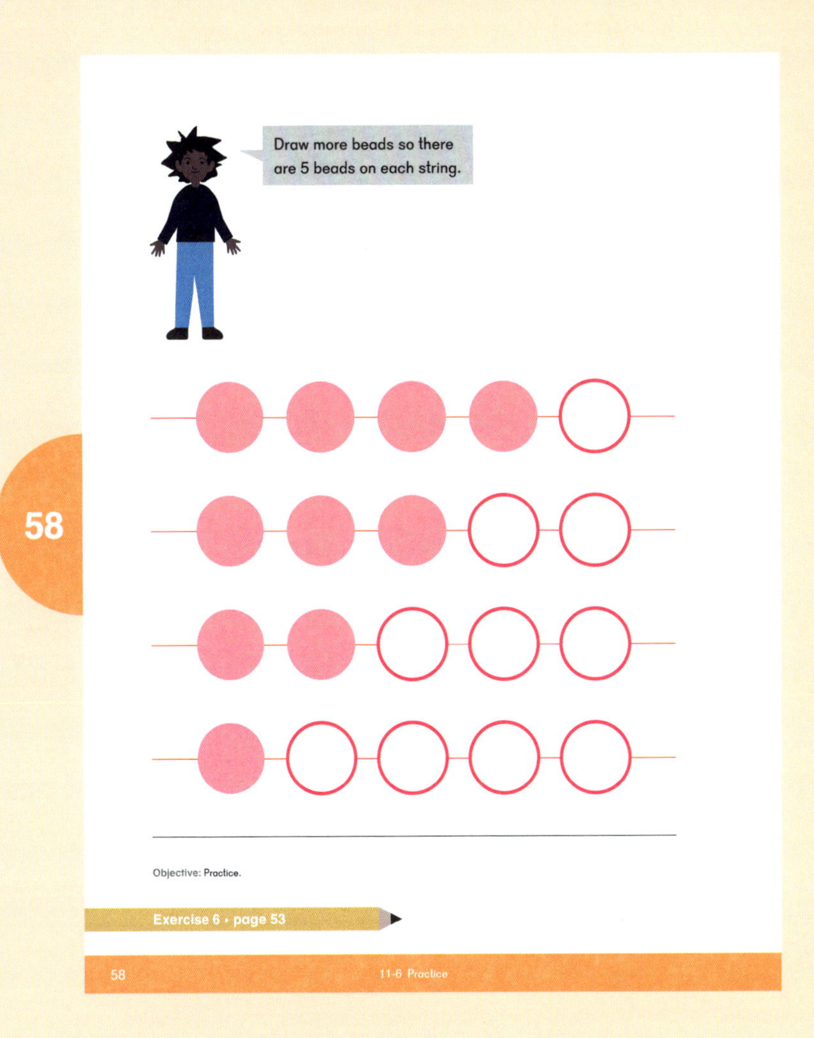

Exercise 1 • pages 39–40

Chapter 11 Compose and Decompose

Exercise 1

1 horse is eating hay.
3 horses are not eating hay.
How many horses are there altogether?

Before using this page: Pre-cut the pictures of horses from cut-outs at back of the book.
Using this page: As you read the number story to students, have them place the same number of horse pictures on the five-frame, then paste the cut-out pictures of horses to show the total.
Concept: Composing four with pictures.

11-1 Altogether — Part 1 39

3 polar bears are swimming.
2 polar bears are standing.
How many polar bears are there altogether?

Before using this page: Pre-cut the pictures of polar bears from cut-outs at back of the book.
Using this page: As you read the number story to students, have them place the same number of polar bear pictures on the five-frame, then paste the cut-out polar bears to show the total.
Concept: Composing five with pictures.

40 11-1 Altogether — Part 1

Exercise 2 • pages 41–42

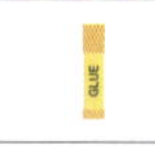

Exercise 2

How many beads are there in all?

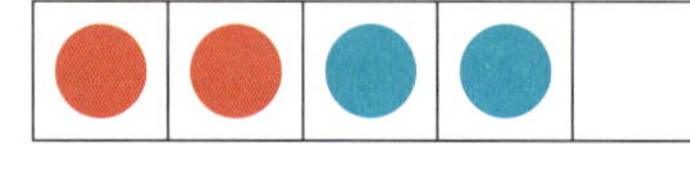

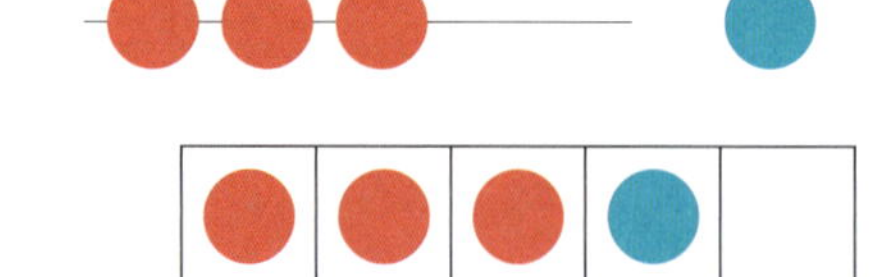

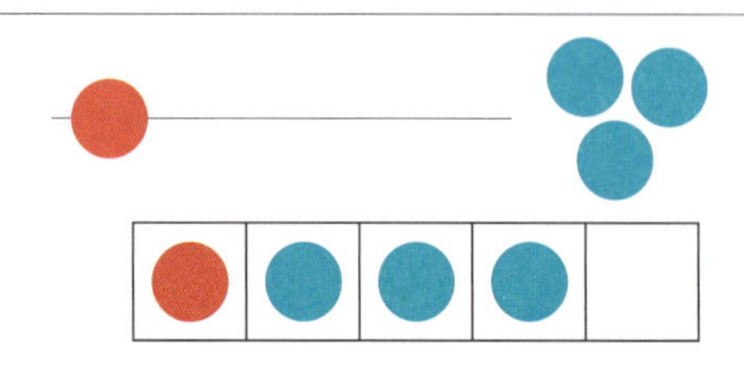

Before using this page: Pre-cut the pictures of beads from cut-outs at back of the book.
Using this page: Have students represent each set of beads by pasting cutouts of the same color and number on the five-frame.
Concept: Composing four with pictures.

11-2 Altogether — Part 2 41

How many beads are there in all?

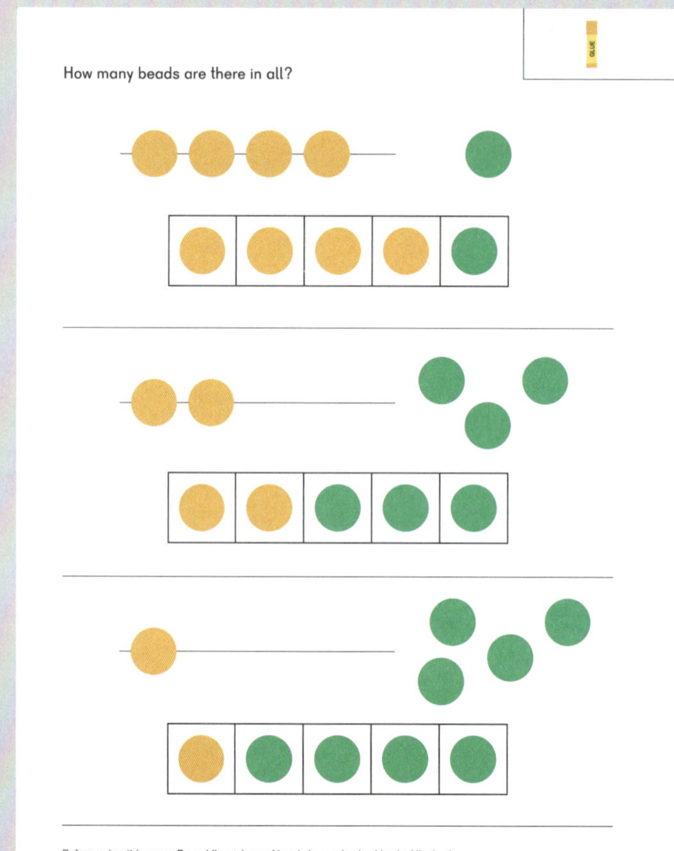

Before using this page: Pre-cut the pictures of beads from cut-outs at back of the book.
Using this page: Have students represent each set of beads by pasting cutouts of the same color and number on the five-frame.
Concept: Composing five with pictures.

42 11-2 Altogether — Part 2

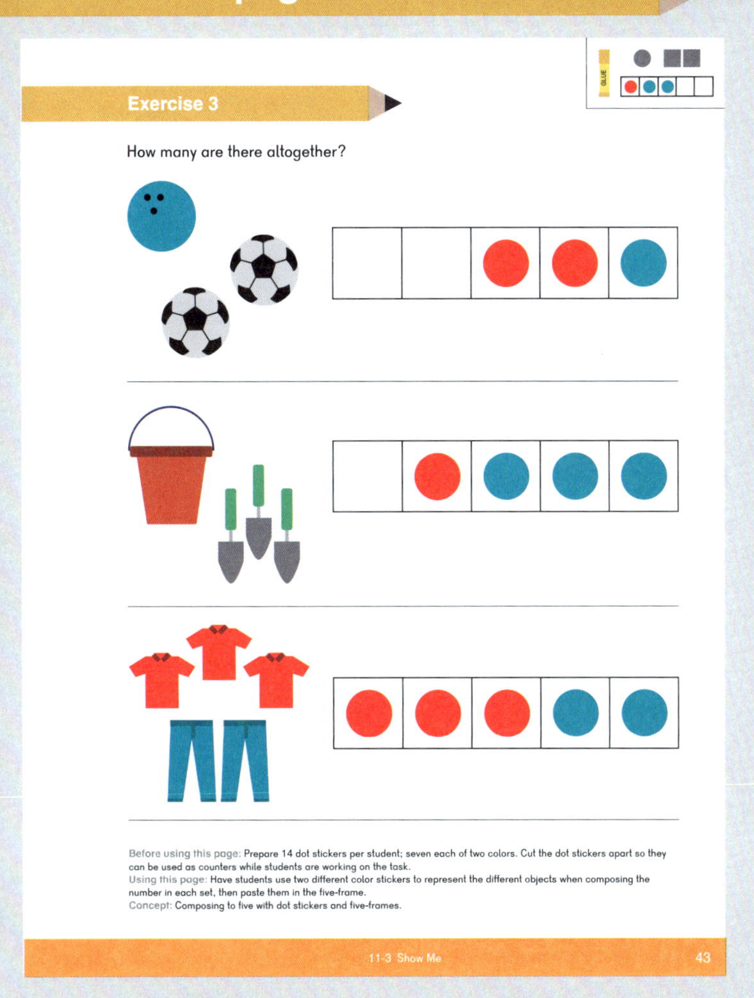

Exercise 3

How many are there altogether?

Before using this page: Prepare 14 dot stickers per student; seven each of two colors. Cut the dot stickers apart so they can be used as counters while students are working on the task.
Using this page: Have students use two different color stickers to represent the different objects when composing the number in each set, then paste them in the five-frame.
Concept: Composing to five with dot stickers and five-frames.

11-3 Show Me 43

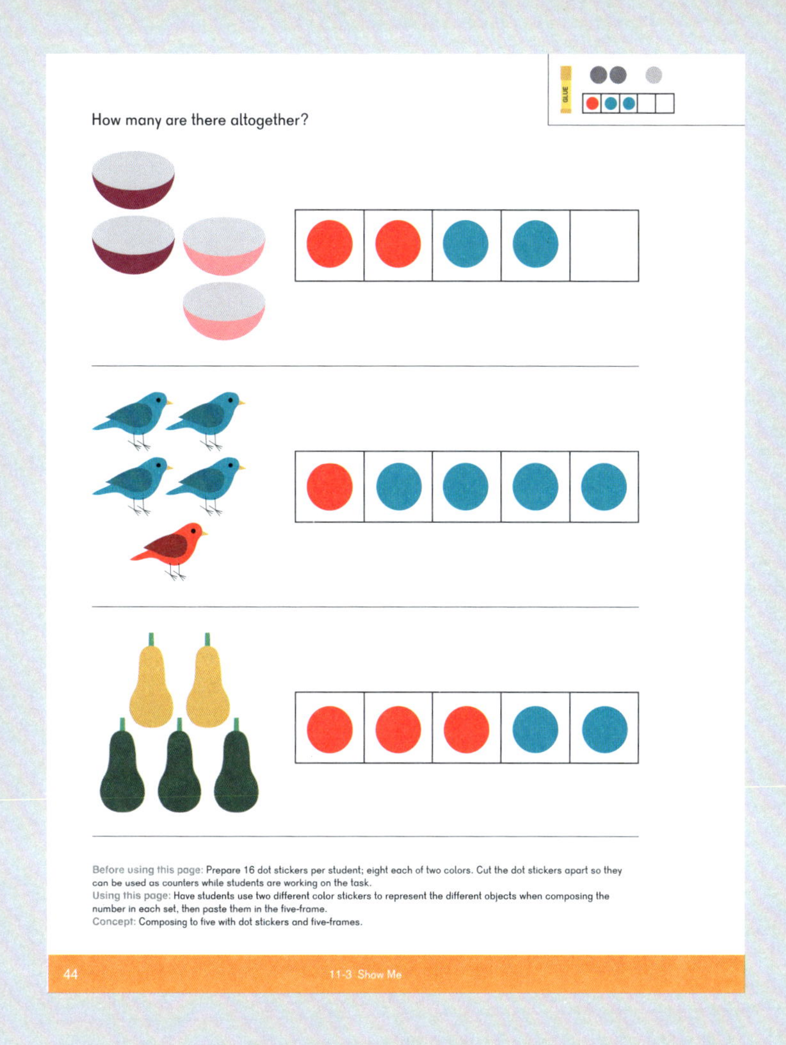

How many are there altogether?

Before using this page: Prepare 16 dot stickers per student; eight each of two colors. Cut the dot stickers apart so they can be used as counters while students are working on the task.
Using this page: Have students use two different color stickers to represent the different objects when composing the number in each set, then paste them in the five-frame.
Concept: Composing to five with dot stickers and five-frames.

44 11-3 Show Me

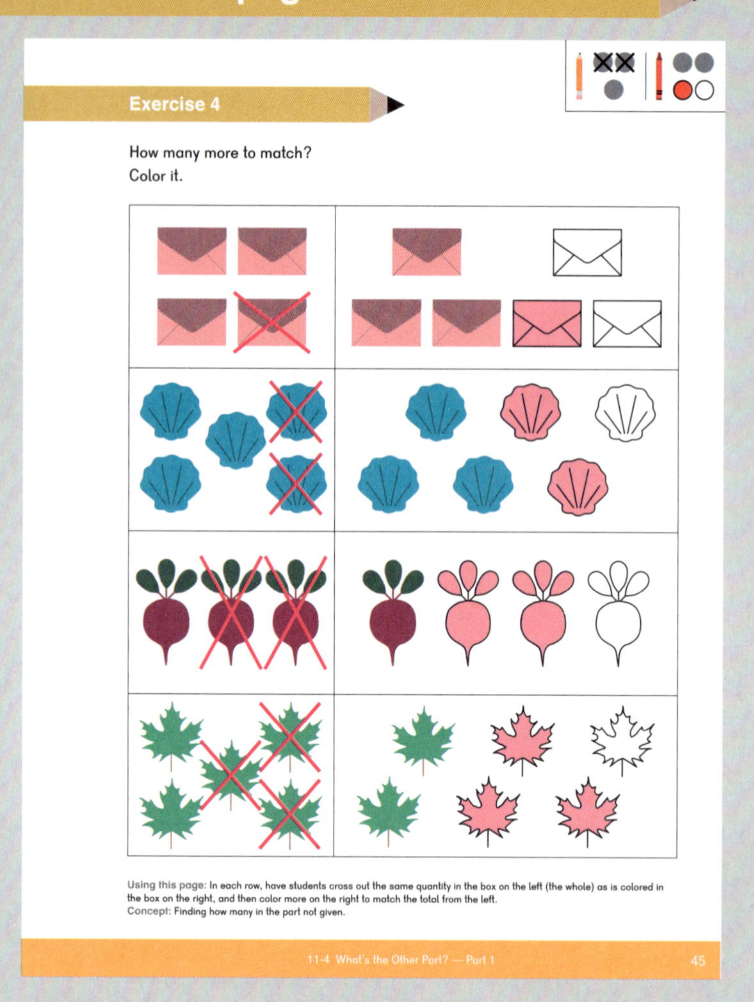
Exercise 4
How many more to match?
Color it.
Using this page: In each row, have students cross out the same quantity in the box on the left (the whole) as is colored in the box on the right, and then color more on the right to match the total from the left.
Concept: Finding how many in the part not given.
11-4 What's the Other Part? — Part 1
45

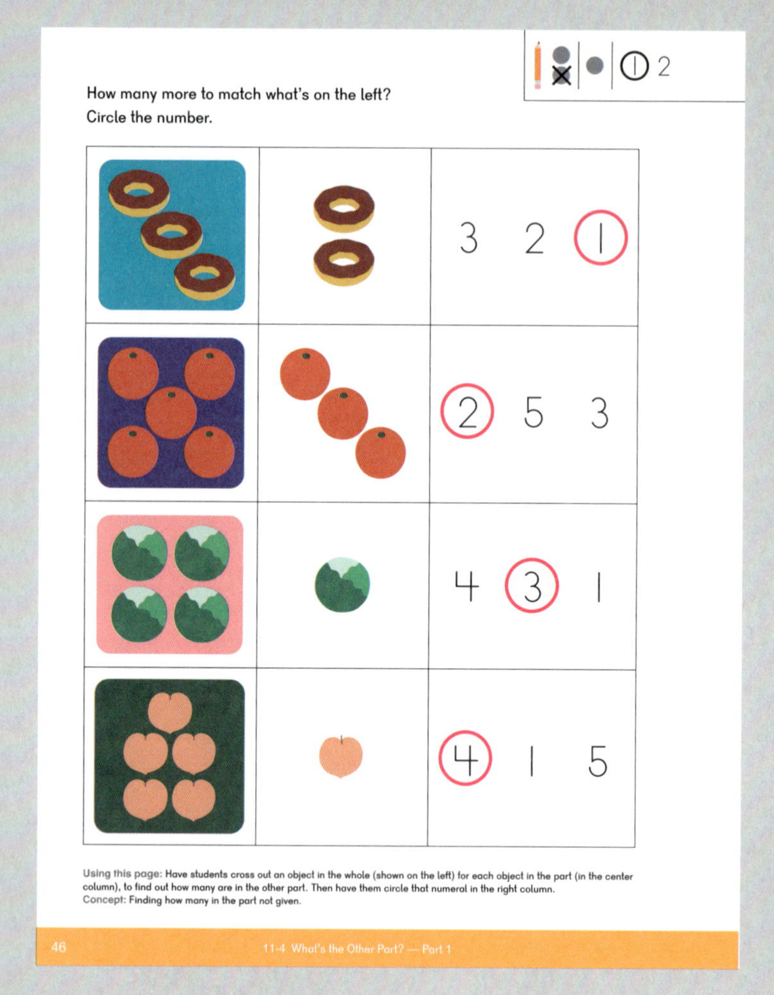
How many more to match what's on the left?
Circle the number.
3 2 1
2 5 3
4 3 1
4 1 5
Using this page: Have students cross out an object in the whole (shown on the left) for each object in the part (in the center column), to find out how many are in the other part. Then have them circle that numeral in the right column.
Concept: Finding how many in the part not given.
46
11-4 What's the Other Part? — Part 1

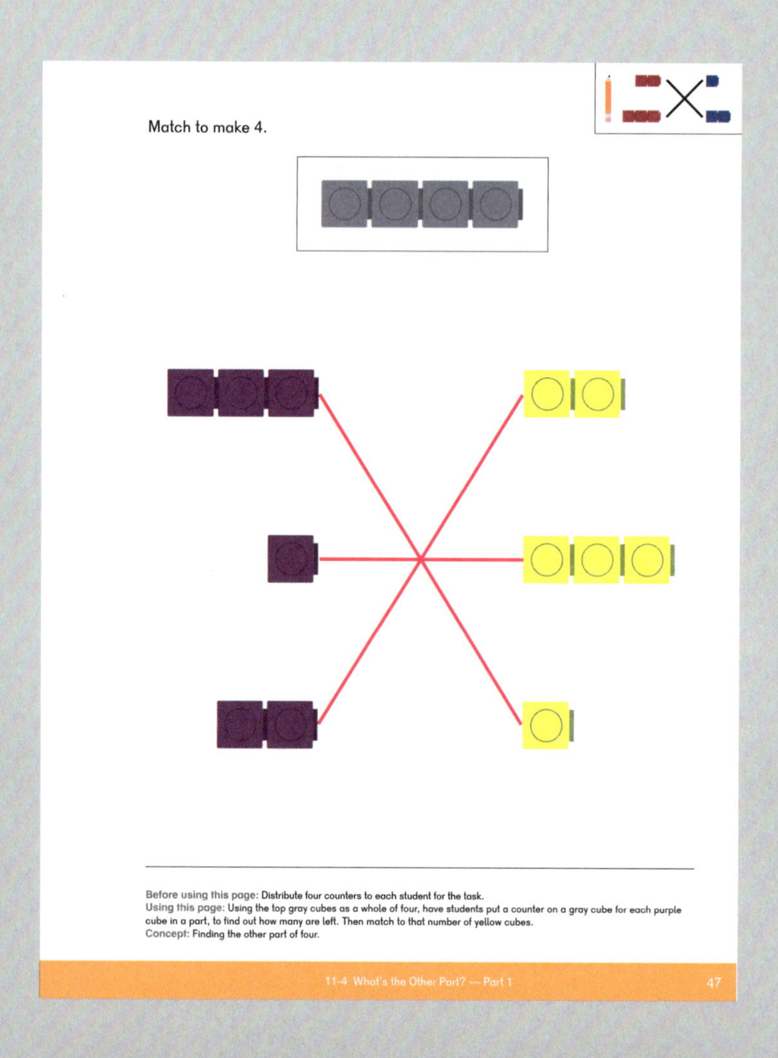
Match to make 4.
Before using this page: Distribute four counters to each student for the task.
Using this page: Using the top gray cubes as a whole of four, have students put a counter on a gray cube for each purple cube in a part, to find out how many are left. Then match to that number of yellow cubes.
Concept: Finding the other part of four.
11-4 What's the Other Part? — Part 1
47

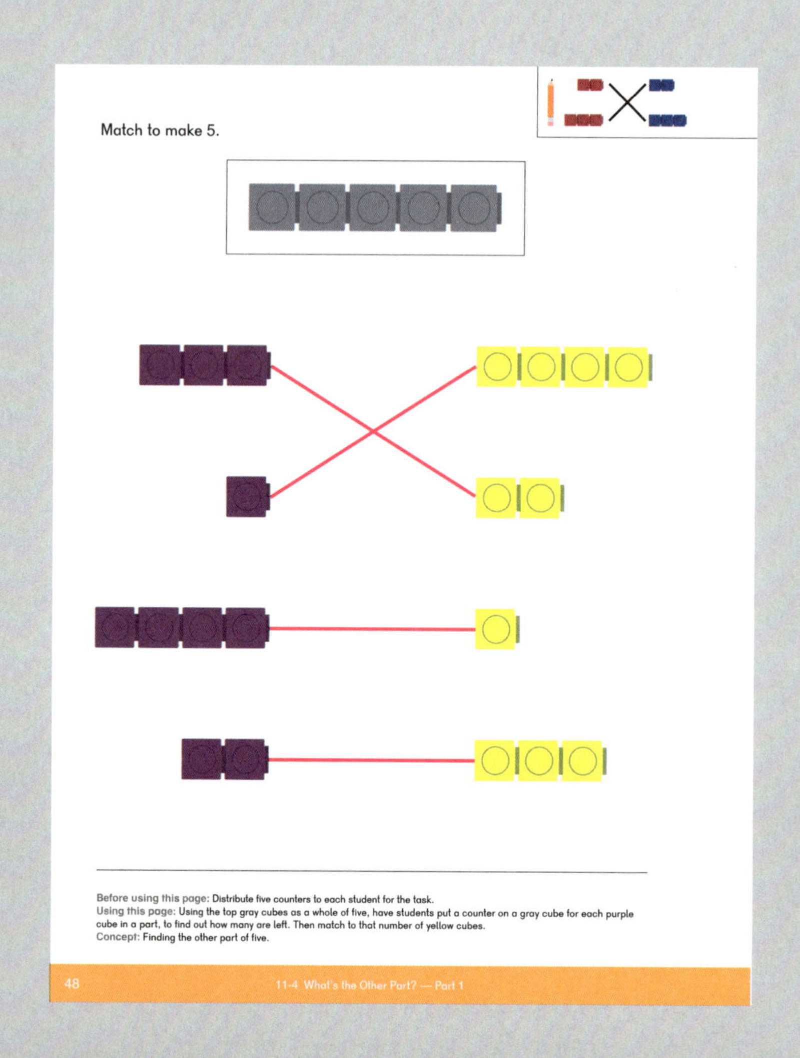
Match to make 5.
Before using this page: Distribute five counters to each student for the task.
Using this page: Using the top gray cubes as a whole of five, have students put a counter on a gray cube for each purple cube in a part, to find out how many are left. Then match to that number of yellow cubes.
Concept: Finding the other part of five.
48
11-4 What's the Other Part? — Part 1

Exercise 5 • pages 49–52

Exercise 5

Each of these ladybugs should have 4 spots altogether on their wings.
Paste the missing spots on the other wing.

Before using this page: Cut apart eight dot stickers for each student.
Using this page: Have students lay out four dot stickers on the right wing of each ladybug, then take away a sticker for each spot on the left wing. Have them paste the stickers that remain on the right wing.
Concept: Finding the other part of four when one part is given.

11-5 What's the Other Part? — Part 2 49

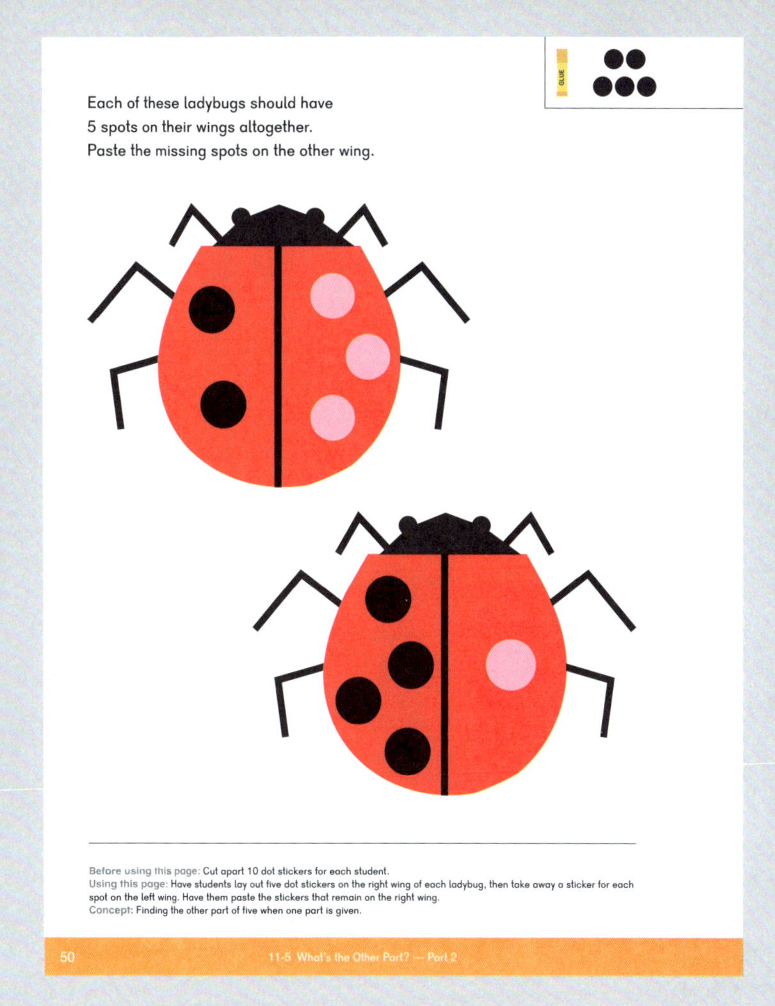

Each of these ladybugs should have
5 spots on their wings altogether.
Paste the missing spots on the other wing.

Before using this page: Cut apart 10 dot stickers for each student.
Using this page: Have students lay out five dot stickers on the right wing of each ladybug, then take away a sticker for each spot on the left wing. Have them paste the stickers that remain on the right wing.
Concept: Finding the other part of five when one part is given.

50 11-5 What's the Other Part? — Part 2

2

How many more to make the number on the right?
Color.

4

5

4

5

Using this page: Have students lay out the same number of counters as the numeral, then take away a counter for each colored square. Have them color the same number of squares as the numbers of counters left.
Concept: Finding the other part within five when one part is given.

11-5 What's the Other Part? — Part 2 51

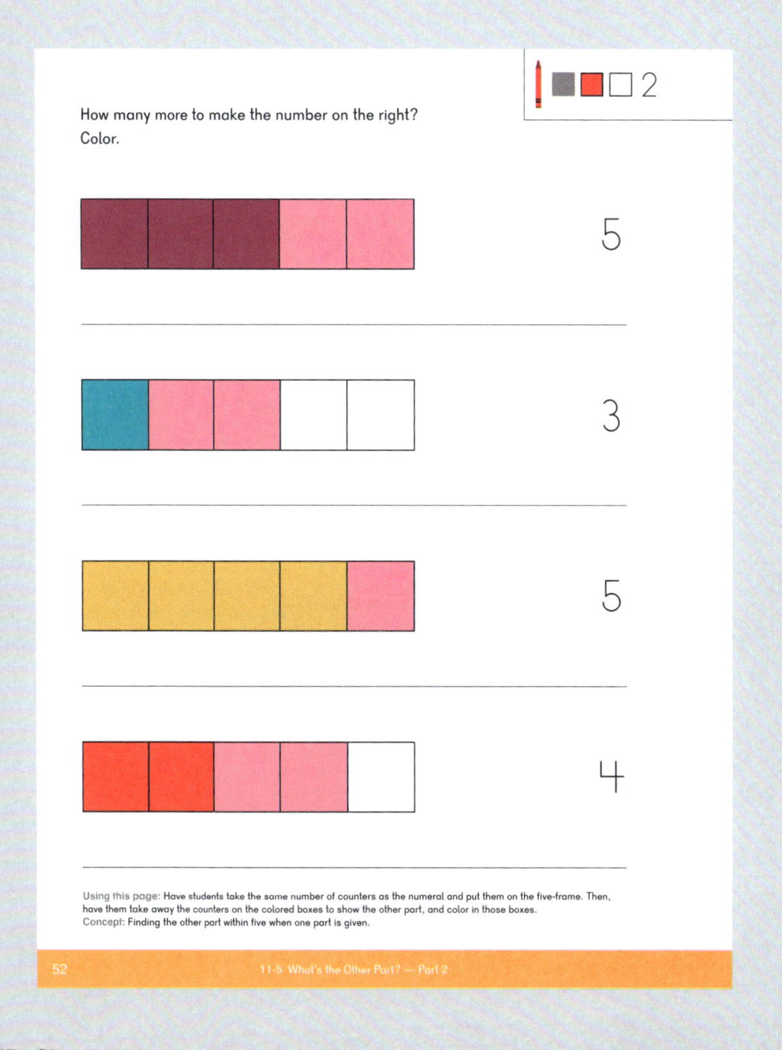

2

How many more to make the number on the right?
Color.

5

3

5

4

Using this page: Have students take the same number of counters as the numeral and put them on the five-frame. Then, have them take away the counters on the colored boxes to show the other part, and color in those boxes.
Concept: Finding the other part within five when one part is given.

52 11-5 What's the Other Part? — Part 2

Exercise 6 • pages 53–54

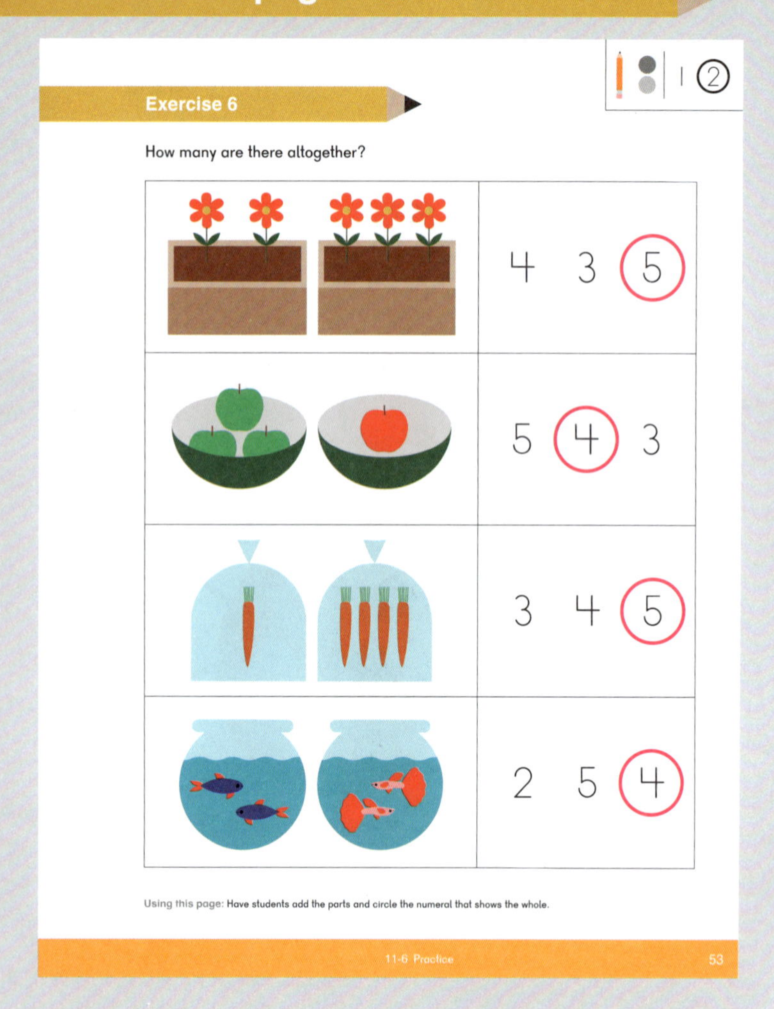

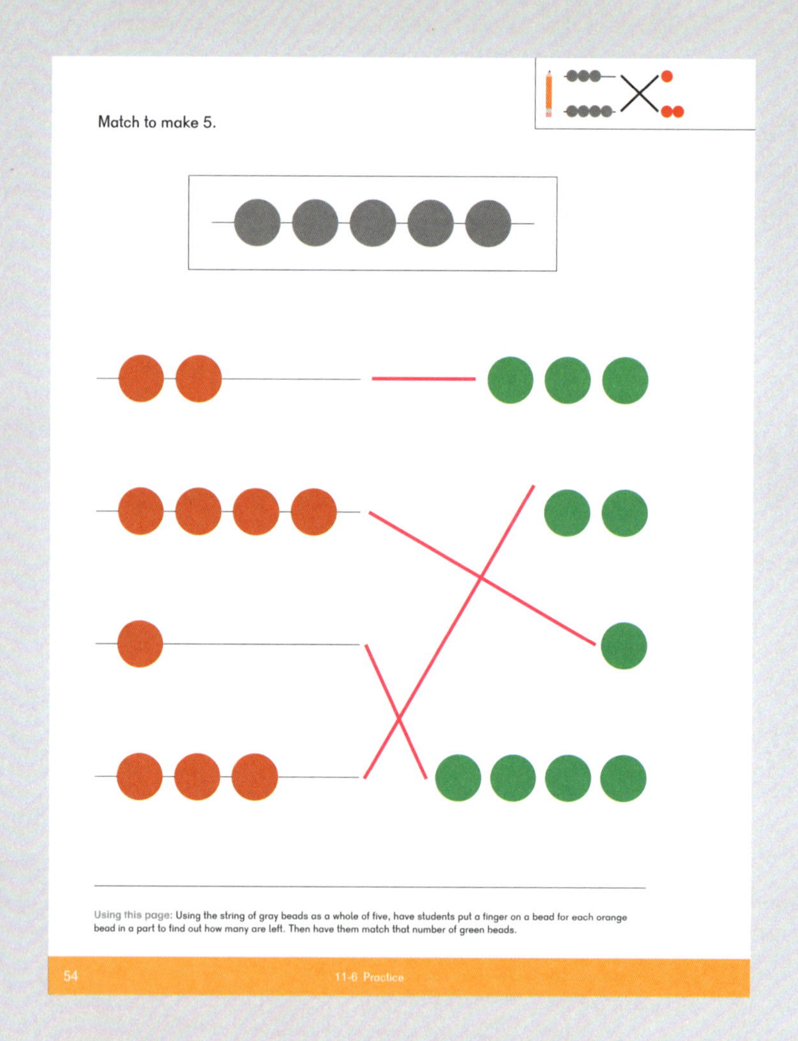

Suggested number of class periods: 8–9

Lesson		Page	Resources	Objectives
	Chapter Opener	p. 91	TB: p. 59	
1	Add to 5 — Part 1	p. 92	TB: p. 60 WB: p. 55	Add numbers to 5.
2	Add to 5 — Part 2	p. 94	TB: p. 61 WB: p. 57	Add numbers to 5.
3	Two Parts Make a Whole	p. 96	TB: p. 63 WB: p. 61	Create addition stories.
4	How Many in All?	p. 98	TB: p. 65 WB: p. 63	Use various strategies to add numbers to 5.
5	Subtract Within 5 — Part 1	p. 100	TB: p. 66 WB: p. 65	Take away numbers within 5.
6	Subtract Within 5 — Part 2	p. 102	TB: p. 68 WB: p. 67	Take away numbers within 5.
7	How Many Are Left?	p. 104	TB: p. 69 WB: p. 69	Use various strategies to subtract numbers within 5.
8	Practice	p. 106	TB: p. 70 WB: p. 71	Practice concepts introduced in this chapter.
	Workbook Solutions	p. 108		

In Chapter 11, students were introduced to composing and decomposing in a very informal manner. In this chapter, students will continue to practice the skills from Chapter 11, using similar and different manipulatives and pictures in order to deepen understanding.

Addition is the simplest of the four operations because it only has one meaning, which is joining. There are two types of addition:

- Static addition is joining two or more parts. In **Dimensions Math® PK**, this type of addition is called "put together."
- Dynamic addition is adding to a part. In **Dimensions Math® PK**, this type of addition is called "adding to."

In PK, only the "put together" or static type of addition will be taught formally. In Kindergarten, students will solve problems using both types of addition. When making up addition stories, be sure that the two parts of the story are already present, and that there is no movement.

Subtraction is a more complex operation than addition because there are three different meanings to subtraction:

- Take away subtraction, which is dynamic
- Part-whole subtraction, which is static
- Comparison subtraction, which is static

In PK, only the take away meaning of subtraction will be taught formally. In Kindergarten, students will solve problems using all three meanings of subtraction. When making up subtraction stories, be sure that there is a change, i.e., that one part is being taken away.

A goal of early childhood math is to have students see the relationship between addition and subtraction. In Chapter 11, students were introduced to that connection through number bonds. Continue to discuss this relationship.

Math research shows that students' willingness to persevere – to have a growth mindset – is incredibly important. We now know that all people can learn math, at any age, and that we learn from making mistakes. The little engine in *The Little Engine that Could* is an example of this type of thinking.

Key Points

The overarching objective of PK math is to provide exposure to numeracy concepts through play so that students' first experiences with mathematics are joyful. Therefore, use of formal terms and symbols should be avoided while teaching this chapter. As you will notice, there are no addition, subtraction, or equality symbols. Even the words "addition" and "subtraction" will not be used with students. Those will all be introduced in Kindergarten.

Differentiate, as necessary, with your students. Allow all students to continue to use objects, fingers, and pictures to solve problems. Encourage visualization throughout the chapter by having students close their eyes when you tell a math story and draw pictures in their minds.

The Commutative Property of Addition will not be taught in PK. This year, "3 and 1" and "1 and 3" are considered two different ways to make 4.

Materials

- Animal stickers
- Bear counters
- Construction paper cutouts of rectangles, squares, circles, and triangles of different sizes and colors
- Containers to be used as vases
- Counters
- Feathers
- Felt bee cutouts
- Felt board
- Fishing poles made in Teacher's Guide PKA Chapter 6, Lesson 3
- Food coloring
- Googly eyes
- Hula hoops
- Index cards
- Jump ropes
- Linking cubes
- Pairs of shoes
- Paper cups
- Paper fish
- Pictures of animals or stuffed toys
- Pipe cleaners
- Plastic eggs or other small objects
- Raisins or grapes
- Raspberries, mini pretzels, or other snacks students can fit over their fingers
- Real or artificial flowers
- Red removable stickers
- Rounded toothpick or coffee stirrers
- Sequins
- Shaving cream
- Spoons
- Tissue paper, cut into 6-in squares

Note: Materials for Activities will be listed in detail in each lesson.

Blackline Masters

- Apple Subtraction Cards
- Apple Tree Template
- Buzzing Bee Template
- Five-frame Cards
- Linking Cube Template
- Number Cards
- Parts and Whole Cards
- Parts Mat Template
- What's Left? Cards

Storybooks

- *The Little Engine that Could* by Watty Piper
- *Little Red Riding Hood* by the Brothers Grimm
- *Do You See Me, Bumble Bee?* by Audrey Muller
- *The Rainbow Fish* by Marcus Pfister

Optional Snacks

- Animal crackers
- Foods from Little Red Riding Hood's picnic basket
- Dumplings made from wonton wrappers or rice cereal treats shaped like dog biscuits
- Finger sandwiches
- Fish crackers
- Orange sections
- Teddy bear crackers

Letters Home

- Chapter 12 Letter

Teddy Bear, Teddy Bear (Lesson 1)

Teddy bear, teddy bear,
Turn around.
Teddy bear, teddy bear,
Touch the ground.
Teddy bear, teddy bear,
Shine your shoes.
Teddy bear, teddy bear, skidoo.
Teddy bear, teddy bear,
Hop in bed.
Teddy bear, teddy bear,
Cover your head.
Teddy bear, teddy bear,
Turn out the light.
Teddy bear, teddy bear,
Say good night.

Teddy bear, teddy bear,
1 and 1
Teddy bear, teddy bear
Makes 2 – this is fun.
Teddy bear, teddy bear
2 and 1
Teddy bear, teddy bear
Makes 3 – so much fun!

Five Little Honey Bees (Lesson 5)

Five little honey bees,
buzzed round the old oak tree
Making some honey in a hive, yum, yum!
One bee then flew away,
up, up and far away.
Now there's four little honey bees. Buzz, buzz!

Four little honey bees,
buzzed round the old oak tree
Making some honey in a hive, yum, yum!
One bee then flew away,
up, up and far away.
Now there's three little honey bees. Buzz, buzz!

Three little honey bees,
buzzed round the old oak tree
Making some honey in a hive, yum, yum!
One bee then flew away,
up, up and far away.
Now there's two little honey bees. Buzz, buzz!

Two little honey bees,
buzzed round the old oak tree
Making some honey in a hive, yum, yum!
One bee then flew away,
up, up and far away.
Now there's one little honey bee. Buzz, buzz!

One little honey bee,
buzzed round the old oak tree
Making some honey in a hive, yum, yum!
One bee then flew away,
up, up and far away.
Now there are no little honey bees. Buzz, buzz!

Lesson Materials

- Counters, 3 per student
- Paper cups, 2 per student
- *The Little Engine That Could* by Watty Piper

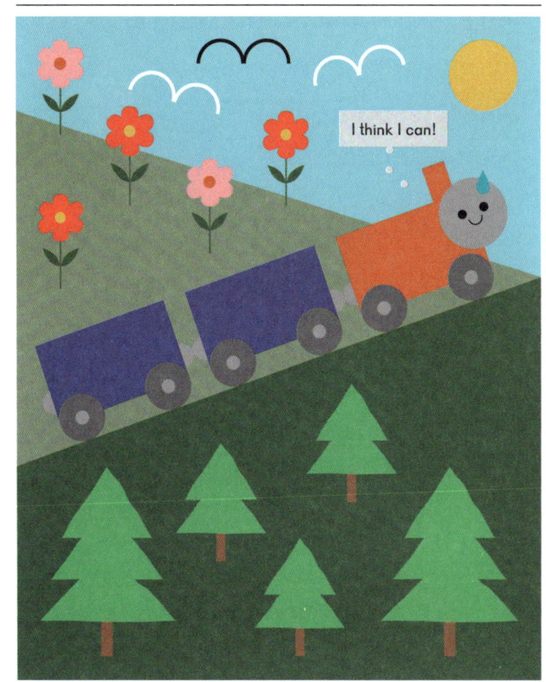

Explore

Give each student paper cups and counters. Tell them that they are to pretend that the paper cups are train cars and the counters are people. Then tell them that three people are riding the train. Have them put counters in the cars to show how many people might be in each car. Ask them if there could be a different way that the people might be riding the train.

Learn

Discuss student solutions to **Explore**. Draw pictures of their solutions for them to see, each time having them say, "____ and ____ make 3."

As students offer their ideas, record in number bonds. For variation, rather than drawing circles for the parts and the whole, you may choose to draw rectangles and have them look like train cars.

Have students look at page 59. Make up addition stories about the objects in the picture and have students model the stories with their counters. Have them say, "____ and ____ make ____" after they model each story.

Extend Learn

Reading Time: Read *The Little Engine That Could* by Watty Piper to the students. Read it again, and have students pretend to be the little engine, huffing and puffing while saying, "I think I can." Discuss the importance of perseverance.

Objective

- Add numbers to 5.

Lesson Materials

- Parts and Whole Cards (BLM)
- Bear counters
- Shaving cream
- Food coloring
- Jump ropes
- Optional snack: teddy bear crackers

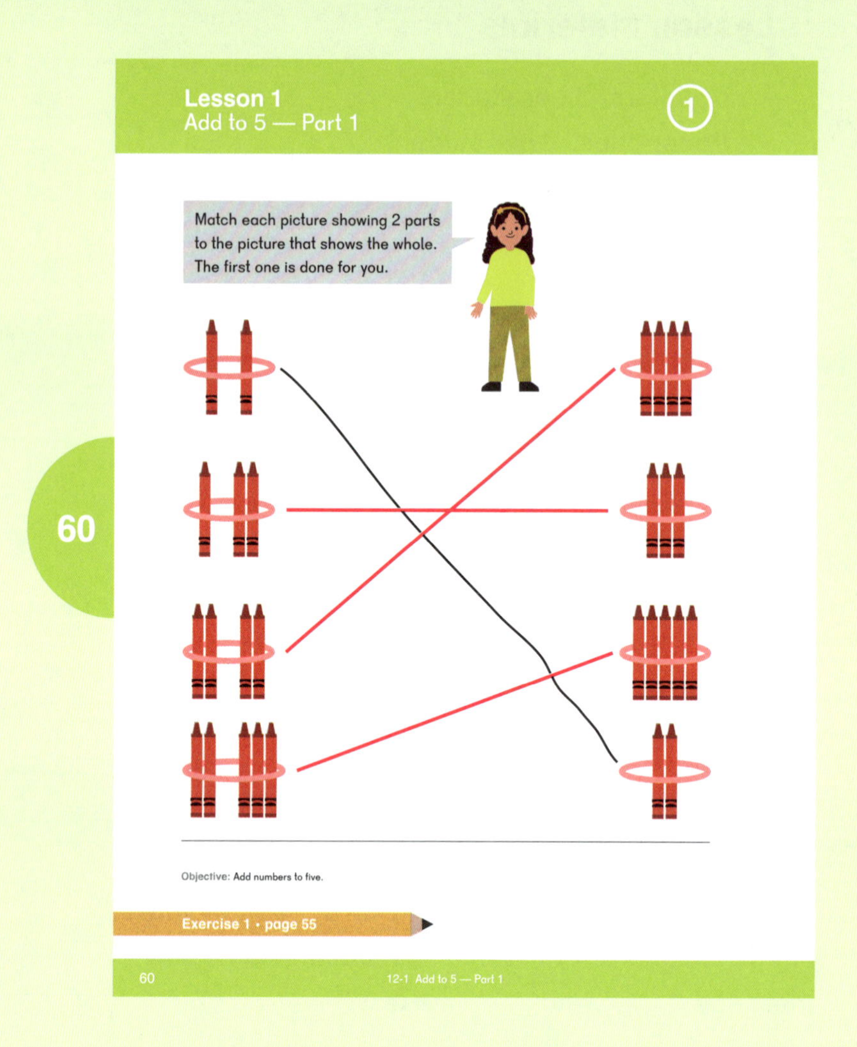

Explore

Give each student either a Parts Card or a Whole Card from Parts and Whole Cards (BLM). Set out bear counters for students to use, if necessary. Have each student find a partner. For example, a student with a Parts Card showing 2 bears and 1 bear would find a student with a Whole Card showing 3 bears.

After all students have found their partners, have students with the same wholes sit together, and use bear counters to model the parts on their cards. Call on several students and have them tell the different ways their group found to make 2, 3, etc. You might differentiate this by giving the smaller numbers to students who are still developing their part/whole understanding and the larger numbers to more confident learners.

Learn

Have students look at page 60. Read Sofia's speech bubble. Have them complete the task.

Whole Group Play

Teach students "Teddy Bear, Teddy Bear" from page 90. After they are familiar with the rhyme, have them do the actions. In the third verse, have them use fingers on both hands to show the parts, then put their hands together to show the whole.

Small Group Center Play

Shaving Cream Parts: Provide shaving cream and food coloring. Have students mix food coloring into shaving cream to create colors of their choice. Give each student a piece of art paper folded in half and tell them that the paper is showing two parts. Have them use two colors of shaving cream, one on each side of the paper, to create art masterpieces. The addition sentence for them to say when done is, "1 and 1 make 2." They should also say, "Part (pointing at one side of the paper) and part (pointing at the other side of the paper) make whole."

Jump Rope: Have students recite "Teddy Bear, Teddy Bear" as they jump over the rope. Play outside if possible.

Exercise 1 • page 55

Extend Learn

Have students draw pictures, using two colors, of the different ways they found to make 2, 3, 4, or 5.

Objective

- Add numbers to 5.

Lesson Materials

- Students' shoes and extra pairs of shoes
- 2 hula hoops of different colors
- Linking cubes, 5 per student
- Parts Mat Template (BLM)
- Parts and Whole Cards (BLM)
- Construction paper cutouts of rectangles, squares, circles, and triangles of different sizes and colors
- Linking cube towers of 4 or 5, in pairs
- Number Cards (BLM) 1 to 3, 1 set per pair of students
- Counters
- Optional snack: dumplings made from wonton wrappers or rice cereal treats shaped like dog biscuits

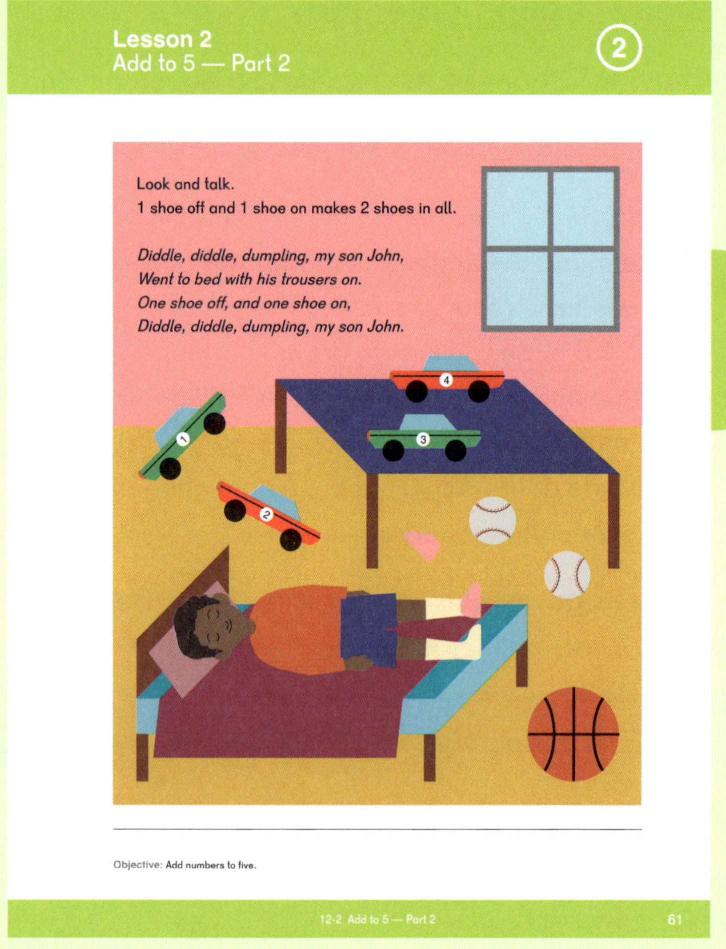

61

Explore

Have students remove their shoes. Take the shoes and mix them in piles on the floor. Have students work together to pair the shoes and line up the pairs along a wall. When they have finished lining up the shoes, have them check the line of shoes to be sure that each shoe has a match.

Learn

Have each student bring their shoes to the group and set them in front of them. Give each student a Parts Mat Template (BLM) and a set of linking cubes. Put the hula hoops on the floor in the middle of circle. Pick up 1 shoe belonging to one child and put it in one hula hoop, then pick up 1 shoe belonging to a different child and put it in the other hula hoop.

Ask, "How many shoes are in the ___ (color) hula hoop? That part has 1. Put 1 linking cube on your mat in one of the circles to represent the 1 shoe."

Point to the other hula hoop and say, "How many shoes are in the ___ (color) hula hoop? That part also has 1. Put a linking cube on your mat in the other circle to represent that shoe. How many shoes are there in all? The whole has 2." Have them say the number sentence, "1 and 1 make 2."

Repeat with other numbers of shoes so the whole is within 5, having students use linking cubes to represent shoes on their Parts Mat Templates (BLM). Have them say the number sentence.

Help students put their shoes back on.

Have students look at page 61 and discuss the picture. Read them the nursery rhyme. Ask them questions about the picture, such as, "How many green cars are in John's room? How many red cars are in John's room? How many cars are in John's room altogether?" Have students use linking cubes to model each story and end each story with, "___ and ___ make ___."

Have students look at page 62 and make up number stories about each picture. Read Emma's speech bubble to them and have them complete the task.

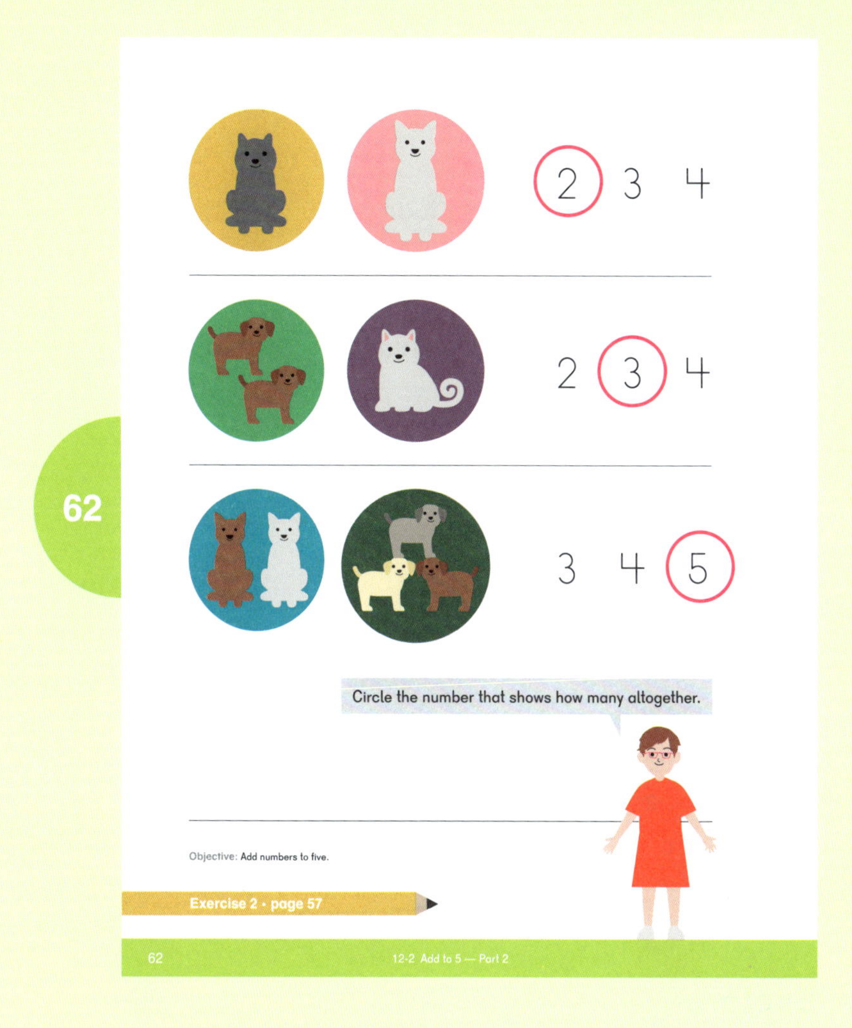

Whole Group Play

Lead students in jumping jacks, windmills, toe touches, etc. Start by telling them, "We will do 1 ___ (exercise). Now we will do 2 ___ (same exercise)." Have them say with you, "1 and 2 make 3." Repeat with other numbers within a whole of 5.

Small Group Center Play

Make a Little Engine: Have page 59 available for students to look at while making a Little Engine. Set out art paper, glue sticks, paper cutouts of rectangles and circles, and crayons. Have students count out 3, 4, or 5 rectangles and circles and create their own train pictures, using two different colors for the cars, gluing them behind the engine, and drawing a face on the engine. When a student's picture is complete, ask him or her to make up a number story about the number of cars of the train, ending with, "___ and ___ make ___."

Matching: Set out a grid of Parts and Whole Cards (BLM) and have students match the Parts Cards to the appropriate Whole Cards.

Snap: Give each pair of students two towers of 4 or 5 linking cubes each, then ask how many cubes are in each tower. Snap one tower into two parts and hide one part behind your back. The other tower will stay on the table as a reference. Have them tell how many cubes are hiding, using the tower on the table to help them if necessary. Then have pairs of students play the game together.

I Wish I Had 4: Set out counters and Number Cards (BLM) 1 to 3, facedown. Playing in pairs, Player 1 will say, "I wish I had 4 but I only have ..." as he turns over a number card and says the number on the card. Player 2 may use counters or fingers, if necessary, to tell how many more are needed to get to 4. Then both students use counters to model the two parts of 4. Students take turns being the Wisher and the Teller.

Exercise 2 • page 57

Objective

- Create addition stories.

Lesson Materials

- Pipe cleaners, 2 per student
- Blue linking cubes, 5 per student
- Pictures of animals or stuffed toys
- Linking cubes, 10 per student, 5 each of 2 different colors and crayons of the same colors
- Linking Cube Template (BLM)
- Animal stickers on index cards, 1, 2, or 3 stickers per card
- 4 containers
- Plastic eggs or other small objects
- 2 spoons
- Number Cards (BLM) 1 to 5
- Counters
- Optional snack: orange sections

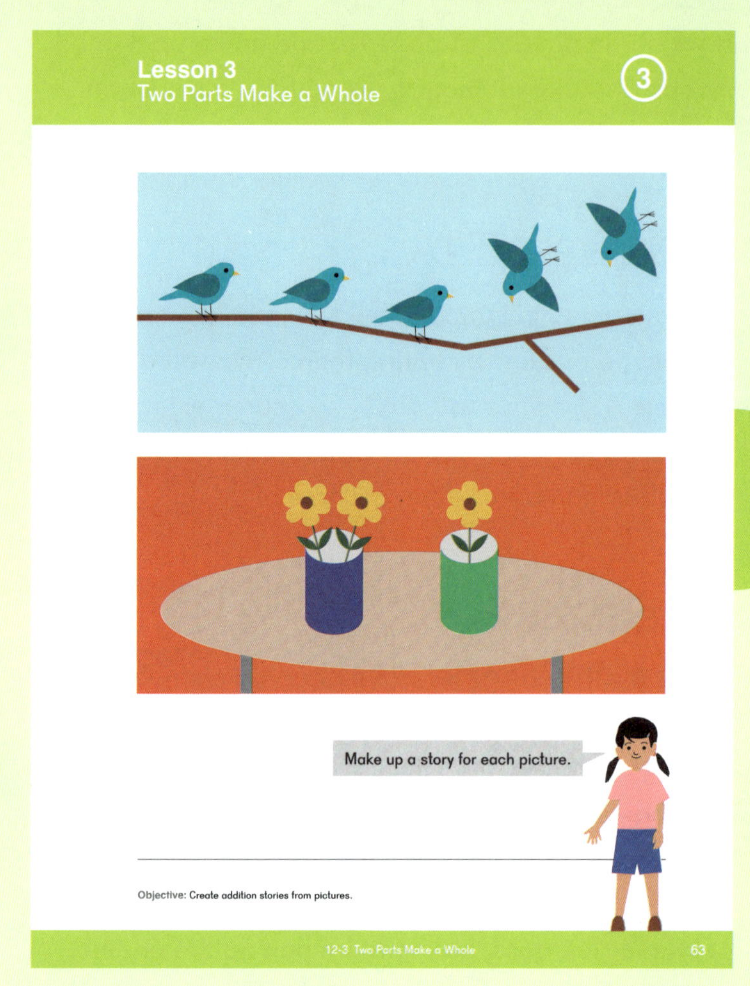

63

Explore

Give each student 2 pipe cleaners and 5 blue linking cubes. Tell them that they are to pretend that the pipe cleaners are tree branches and the cubes are birds. Then tell them that 4 birds are on 2 branches. Have them put cubes on the pipe cleaners to show how many birds might be on each branch. Then have them tell a story to a partner about birds on branches using linking cubes and pipe cleaners. Ask them if there could be different ways that the birds could be on the branches. Have them tell new stories each time they make different parts.

Learn

Hold up 2 pictures of animals, or stuffed toys, such as 1 elephant and 2 elephants for the students to see. Make up an addition story about the animals, such as, "At the zoo, there was 1 elephant eating and 2 elephants not eating. There were 3 elephants in all." Emphasize that 1 and 2 are the parts in the story and 3 is the whole. Tell the story again, this time having students use the fingers on one hand to show 1 part and the fingers on the other hand to show the other part, then put their hands together to show the whole.

Have students look at page 63 and discuss the picture. Read Mei's speech bubble to them. Have them make up stories for each picture, identify the parts and whole for each story, and end with, "___ and ___ make ___."

Have students look at page 64 and identify the objects on the page. Have them make up stories for each picture, and tell their stories to a partner. Have some of them share their stories with the group. For the shared stories, have students identify the parts and whole for each story, and end with, "___ and ___ make ___." Read Dion's speech bubble to them and have them complete the task.

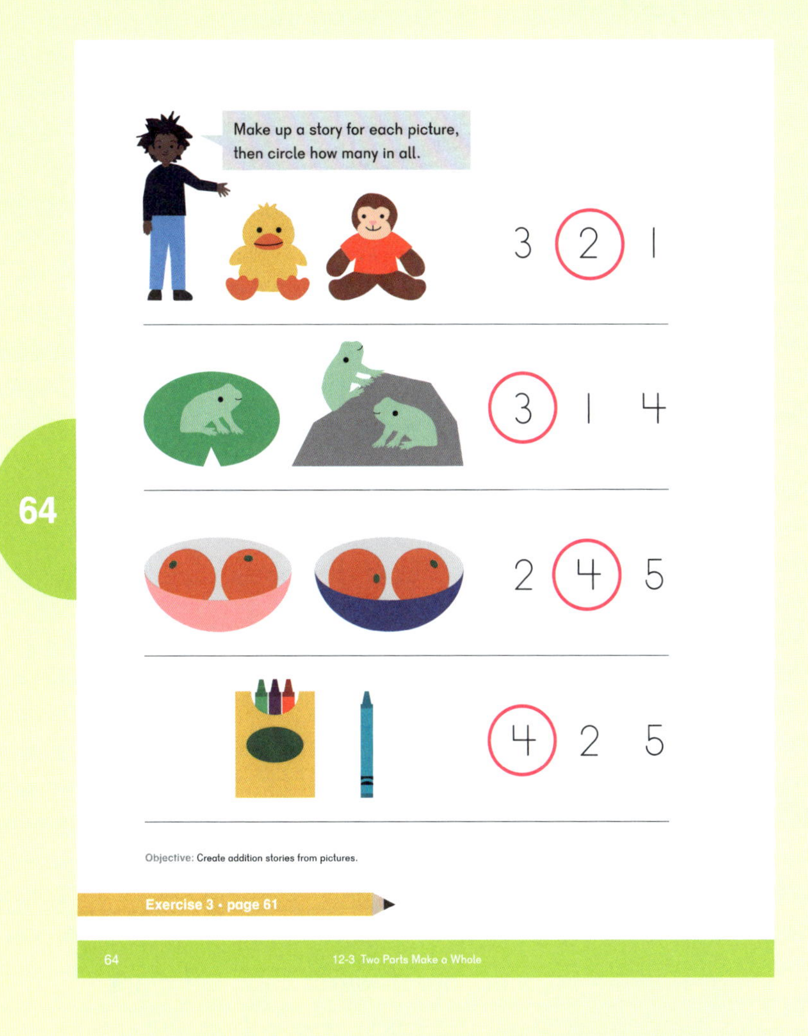

Whole Group Play

Relay Race: Divide students into 2 teams and have them line up single file behind a start line. Place 1 container of plastic eggs for each team at the start line, and one empty container per team at the finish line. Give the first student of each team a spoon. When you say, "Go!" that student uses the spoon to scoop a plastic egg out of the container, and moves as quickly as possible without dropping the egg to the finish line. When the student reaches the finish line, she drops the egg into the container, then moves quickly back to the start line to give the spoon to the next student in line. The team who finishes first wins. When the race is over, pick up the winning team's container and remove the eggs, one at a time. When you remove the second egg, say, "1 and 1 make 2," and have students repeat. When you remove the next object, have students say the number sentence, "2 and 1 make 3." Continue until 5 eggs have been counted.

Small Group Center Play

Tell Me a Story: Set out the animal sticker cards facedown. Have students choose 2 cards and make up stories similar to those told earlier. They can either record their stories or tell them to friends.

Color Your Towers: Set out linking cubes of 2 different colors and crayons of the same colors. Give each student linking cubes and a Linking Cube Template (BLM). Tell students that they have five minutes to use 2 colors of linking cubes to make towers of 5 in as many ways as they can.

Tell students that cubes with the same color need not be adjacent. This should lead to good discussion. For example, 3 and 2 make 5 even when a pattern of red, yellow, red, yellow, red is used.

After five minutes, have them color their templates using the appropriate colors to represent all of the towers they made. Each time they make a tower, they must color one of the linking cube towers on their template using the appropriate 2 colors.

I Wish I Had 5: Play as in the previous lesson, but with a whole of 5.

Exercise 3 • page 61

Lesson 4 How Many in All?

Objective

- Use various strategies to add numbers to 5.

Lesson Materials

- Counters, 5 per student
- Paper cups, 2 per student
- *Little Red Riding Hood*
- Real or artificial flowers, 5 each of 2 different colors
- Containers to be used as vases
- Tissue paper, cut into 6-in squares
- Pipe cleaners
- Optional snack: some of the foods Little Red Riding Hood brought to her grandmother

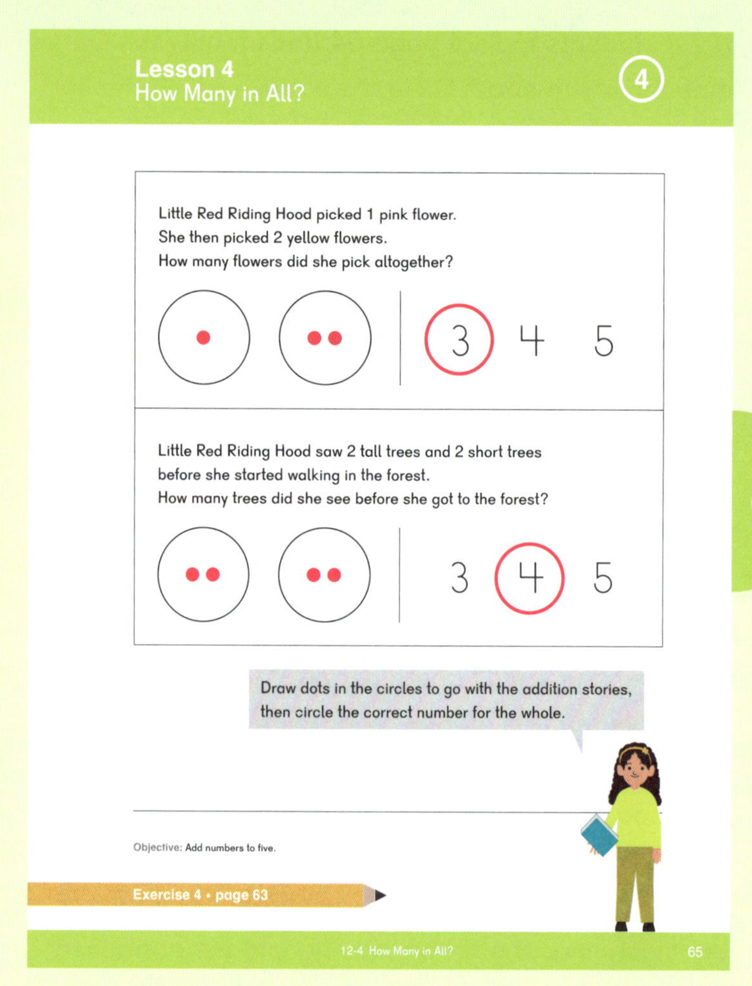

Explore

Give each student paper cups and counters. Tell them that they are to pretend that the paper cups are vases and the counters are flowers. Then tell them that five flowers are in the vases. Have them put "flowers" in the "vases" to show how many flowers might be in each vase. Have them tell a partner a story about flowers in vases using the counters and cups. Ask them if there could be a different way that the flowers could be in the vases. Have them tell a story each time they make different parts. As they tell their stories, use real or artificial flowers and vases to model the stories they tell.

Learn

Ask students what they might bring to an elderly person who is not feeling well. If necessary, suggest that bringing flowers often helps to cheer people up.

Read *Little Red Riding Hood*. Have students recall what Little Red Riding Hood was bringing to her grandmother. Then ask students what they think the little girl's favorite color is. Ask them if they remember which character has the same favorite color (Emma). Have students look at page 65. Read Sofia's speech bubble to them and have them complete the task.

Whole Group Play

Lead students in jumping jacks, windmills, toe touches, etc. Start by telling them, "We will do 1 ___ (exercise). Now we will do 2 ___ (same exercise)." Have them say with you, "One and two makes three." Repeat with other numbers within a whole of 5.

Small Group Center Play

Dramatic Play: Have students dress up like characters from *Little Red Riding Hood* and act out the story.

Tissue Paper Flowers: Cut tissue paper of different colors into 6-inch squares. Have each child choose 3 paper squares and fold them as if they were creating a fan. Wrap a pipe cleaner around the middle of the folded paper and twist it so that it holds the paper together. Have students carefully separate the pieces of paper.

Exercise 4 • page 63

Extend Learn

Book Talk: Ask students what lessons could be learned from *Little Red Riding Hood*. For example, ask them if the little girl should have been talking to a stranger.

Lesson 5 Subtract Within 5 — Part 1

Objective

- Take away numbers within 5.

Lesson Materials

- Felt board
- 5 felt bee cutouts
- Paper cups, 1 per student
- Yellow linking cubes
- Raspberries, mini pretzels, or other snacks students can fit over their fingers, 5 per student
- Buzzing Bee Template (BLM)
- Number Cards (BLM) 1 to 5
- What's Left? Cards (BLM)
- Linking cube towers of 5, in pairs
- *Do You See Me, Bumble Bee?* by Audrey Muller
- Optional snack: snack materials used in **Whole Group Play**

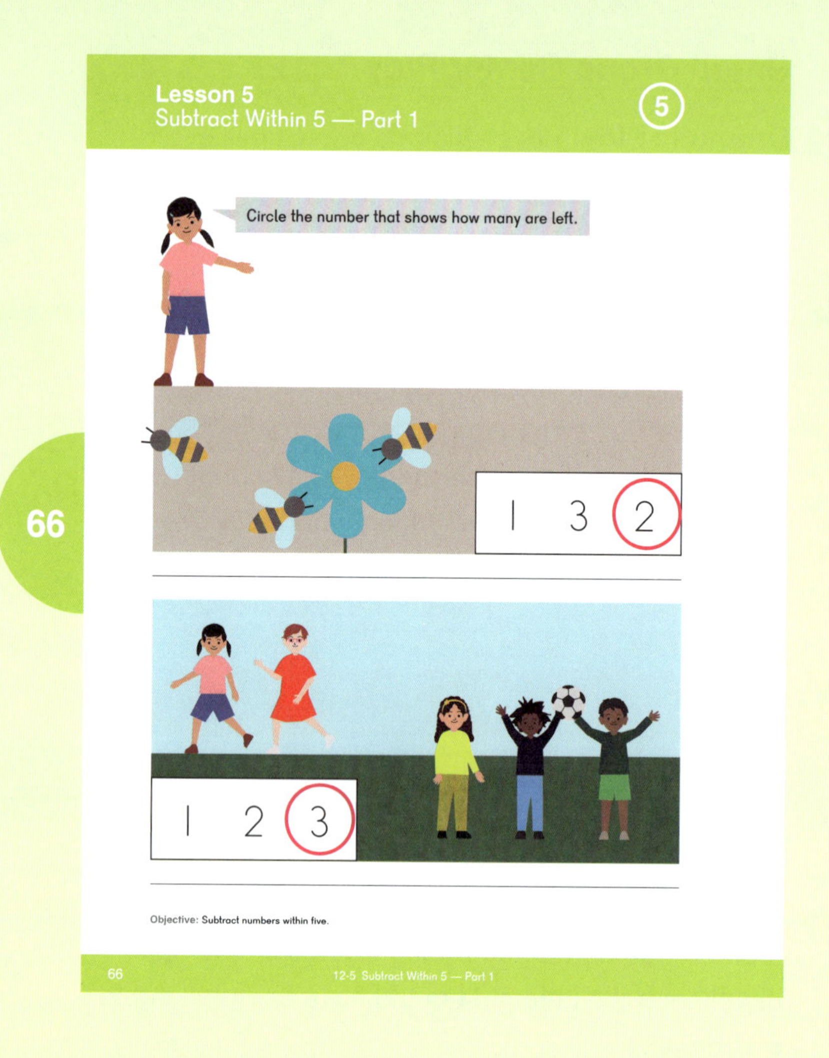

Explore

Teach students the song "Five Little Honey Bees," sung to the tune of "Five Little Speckled Frogs." Use the felt bee cutouts and the felt board as you teach the song.

As you sing the song, have students pretend that the yellow cubes are bees. As each bee flies away, have them put the cube into the paper cup. The last line of the first verse would then be, "Now there's 4 little honey bees. Buzz, buzz." Repeat, having 2 more bees fly away. When you get to the last line, pause before saying how many bees are left and let students help you. Sing one more time so that there are no bees left.

Learn

Have students tell stories about up to 5 honey bees with some of the bees leaving, and have the other students act out the stories. As some students are acting, other students will model the story using their linking cubes. After a student tells a story, ask questions. Encourage students to answer the questions in complete sentences.

3 bees are buzzing around a flower. 1 of the bees has to go back to the hive. Ask:

- How many bees were buzzing around the flower at first?
- How many bees went back to the hive?
- How many bees are still buzzing around the flower? (There are 2 bees still buzzing around the flower.)

Be sure that at least one story has 0 bees left.

Have students look at page 66 and discuss the pictures by telling stories and asking questions similar to those above. Read Mei's speech bubble and have students complete the task.

Have students look at page 67. Repeat the procedure on page 66.

Whole Group Play

Give each student 5 raspberries or other snacks they can fit over their fingers. Have them put them on the fingers of one hand. Tell subtraction stories about eating the snacks and have the students act them out. For example, "I woke up this morning with 5 ___ (name of snacks) on my fingers. I ate 1 ___. How many ___ were left on my fingers?" (There were 4 ___ left on your fingers.) "Later in the morning, I had 4 ___ on my fingers. I ate 2 ___. How many ___ were left on my fingers?" (There were 2 ___ left on your fingers.) Repeat until all snacks are eaten.

Small Group Center Play

How Many Buzzing Bees are Left?: Set out Buzzing Bee Templates (BLM). Have students choose a template and match it to a Number Card (BLM) to show how many bees are left. Then have students color the template and decorate as desired.

What's Left Story: Give each group of students a What's Left? Card (BLM) and have them act out the story shown in the picture.

Snap With a Twist: Have pairs of students tell each other a "What's Left" story and model the story with linking cubes. Then have them play **Snap**, pretending that the linking cubes are the objects in their stories.

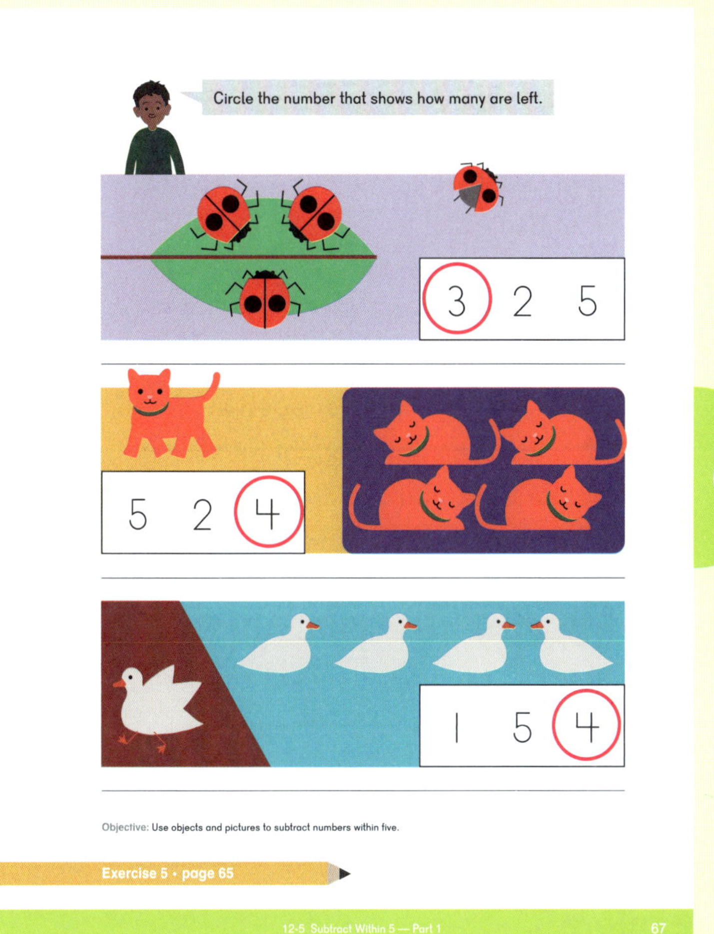

67

Exercise 5 • page 65

Extend Learn

Bee Talk: Either read *Do You See Me, Bumble Bee?* or tell students about the importance of bees to agriculture.

Objective

- Take away numbers within 5.

Lesson Materials

- Felt board
- Felt bee cutouts
- Yellow linking cubes
- Number Cards (BLM) 1 to 5, 1 set per student
- 5 raisins or grapes for each student
- Rounded toothpick or coffee stirrers
- Apple Subtraction Cards (BLM)
- Paper cups, 1 per student
- Linking cube towers of 5, in pairs
- Apple Tree Template (BLM)
- Red removable stickers or counters, 5 per student
- Optional snack: animal crackers

68

Lesson 6
Subtract Within 5 — Part 2

6

Cross out the number of apples that have fallen off the tree.
The first one is done for you.

There were 5 apples on the tree.
1 apple fell on the ground.
How many apples were left on the tree? 4 apples were left.

There were 4 apples on the tree.
2 apples fell on the ground.
How many apples were left on the tree? 2 apples were left.

There were 3 apples on the tree.
1 apple fell on the ground.
How many apples were left on the tree? 2 apples were left.

Objective: Subtract numbers within five.

Exercise 6 • page 67

68 12-6 Subtract Within 5 — Part 2

Explore

Put five bees on the felt board.

Sing "Five Little Honey Bees" again. Change the words to have two bees fly away. Have students work in pairs with linking cubes to imagine what the felt board will look like after the bees fly away. After a couple of minutes, remove the bees from the felt board and have students help you figure out how many are left.

Learn

Give each student a paper cup and 5 raisins or grapes on a rounded toothpick or coffee stirrer. Tell students to pretend that the toothpick or stirrer is a tree branch and the raisins or grapes are apples. Tell a story about 5 apples on a tree branch where 1 fell off. As you tell the story, have students remove a raisin or grape and put it in the paper cup. Ask them how many apples are left on the branch. Encourage them to answer in complete sentences. "There are 4 apples left on the branch." Repeat with 2 apples falling off the branch, then 1, then the last one. Let children eat the fruit.

Give each student a set of Number Cards (BLM). Tell the story again, starting with 4 apples, then 3, then 2, then 1. Instead of saying the number sentences, have students show the Number Card (BLM) that shows the number of apples left. Finally, have students use their fingers, instead of fruit, to model the stories you tell them. The number you subtract from 5 can be 1, 2, or 3. Ensure that students are using their fingers correctly to model the stories you tell.

Have students look at page 68 and discuss the pictures. Ask them how this page is like the song they sang and the activity they did with raisins or grapes. Tell them that one way to show subtraction in pictures is to cross off the objects that are being taken away. Guide them in completing the task.

Whole Group Play

Put students in groups of 5. Have each group hold hands and stand in a circle. Tell take away stories and have the number of students who are being taken away sit down. After each story, have students say, for example, "There were 5 friends standing at first. 2 friends sat down. There are 3 friends left standing."

Small Group Center Play

Apple Number Matching: Set out Apple Subtraction Cards (BLM) and Number Cards (BLM). Have students match what's left on each apple subtraction card to the appropriate numeral card.

Apple Story Time: Set out Apple Subtraction Cards (BLM). Have students tell stories about the cards.

Snap: Up to a whole of 5.

Exercise 6 • page 67

Extend Learn

Comparing Stories: Give each student an Apple Tree Template (BLM) and red stickers or counters. Tell the following stories and have students model them on their templates.

- There were 3 apples on a branch of an apple tree. There were 2 apples on another branch of the apple tree. How many apples were in the tree altogether? (There were 5 apples in the tree altogether.)
- There were 5 apples on a tree. 2 of the apples fell to the ground. How many apples were still in the tree? (There were 3 apples still in the tree.)

Ask students what is alike and what is different between the two stories. You may hear responses such as:

- Both stories were about apples and trees.
- In the first story, I had to put the number of apples together to find the answer. In the second story, I had to take 2 apples away from 5 apples.

Objective

- Use various strategies to subtract numbers within 5.

Lesson Materials

- Counters, 5 per student
- Five Frame Cards (BLM) 1 to 5
- Fishing poles made in Teacher's Guide PKA Chapter 6, Lesson 3
- Paper fish
- Googly eyes
- Feathers
- Sequins
- *The Rainbow Fish* by Marcus Pfister
- Optional snack: fish crackers

Explore

Teach students "Five Little Fishes."

Five Little Fishes

Five little fishes swimming in the sea
If one swims away, how many will there be?

Four little fishes swimming in the sea
If one swims away, how many will there be?

Three little fishes swimming in the sea
If one swims away, how many will there be?

Two little fishes swimming in the sea
If one swims away, how many will there be?

One little fish swimming in the sea
If one swims away, how many will there be?

After children are familiar with the rhyme, have them answer the question at the end of each verse. If necessary, have students act out the rhyme to help them answer the questions and give each of the other students 5 counters.

Learn

Have students look at page 69. Read Mei's speech bubble. Have students tell a story about the fish in the first row, then circle the 3. Repeat for the second and third rows. The fourth row, having 0 objects left, is new in this chapter. It may be helpful for students to use their fingers to model the stories on this page. Tell other stories where there are 0 objects left, starting with different numbers, and have students model the stories with their fingers.

Whole Group Play

I Wish I Had 5 on My Card: Flash Five-frame Cards (BLM) showing a part of 5. Have students hold up fingers to show the other part. Then have students say, "Five is ___ (number you showed) and ___ (number missing.)"

Small Group Center Play

Go Fish: Give each pair of students a fishing pole and 5 paper fish. Have them take turns "fishing." Each time a fish is caught, have them say, "There were ___ fish. I caught 1 fish. There are ___ fish left."

Feather Print Fish: Have students dip a feather in paint and transfer the paint to a piece of paper. After the paint has dried, have students glue googly eyes and sequins onto the fish and draw an underwater scene.

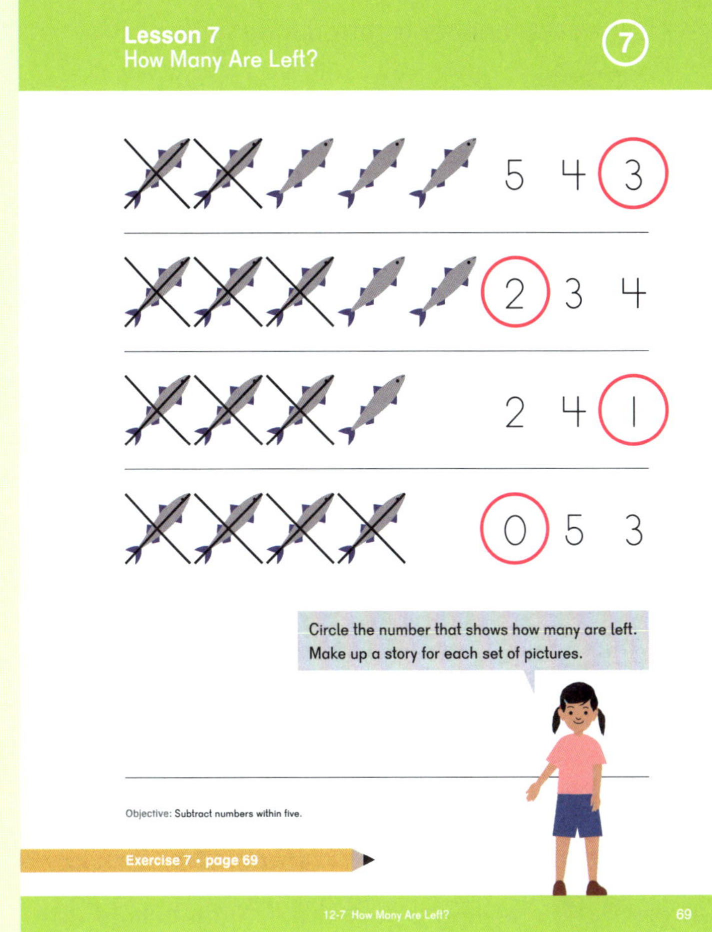

Exercise 7 • page 69

Extend Learn

Reading Time: Read *The Rainbow Fish* to the students and discuss the importance of being oneself.

Objective

- Practice concepts introduced in this chapter.

Lesson Materials

- Five-frame Cards (BLM) for 2 to 5, 1 set
- Number Cards (BLM) 0 to 5, 1 set per student
- Parts and Whole Cards (BLM)
- Apple Subtraction Cards (BLM)
- Optional snack: finger sandwiches

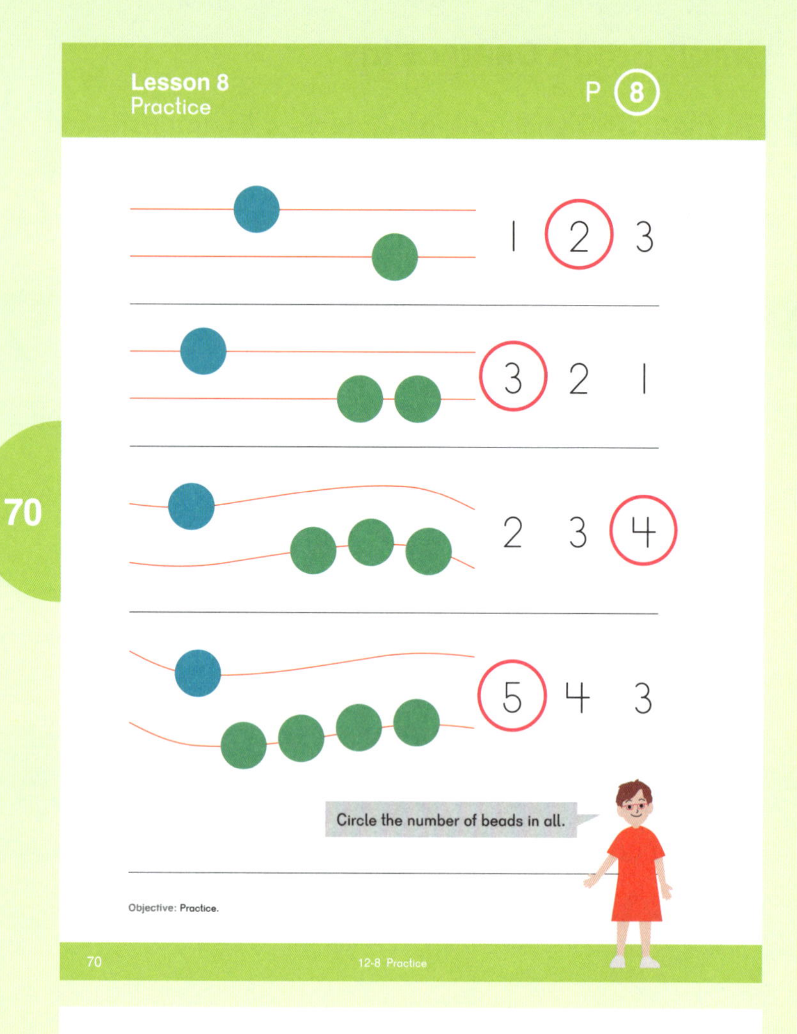

For the **Practice**, read the directions and speech bubbles on each page and have students complete the tasks.

Whole Group Play

I Wish I Had Five on My Card!: Flash Five-frame Cards (BLM) showing a part of 5. Students hold up the Number Card (BLM) showing the other part of 5. Then have students say, "5 is ___ (number you showed) and ___ (number missing)."

Small Group Center Play

Put Together: Using Parts and Whole Cards (BLM), have students match the parts cards to the whole cards.

Matching Subtraction: Set out Apple Subtraction Cards (BLM) and Number Cards (BLM). Have students match what's left on each apple subtraction card to the appropriate numeral card.

Walking Crabs: Have students race from a start line to a finish line doing the crab walk.

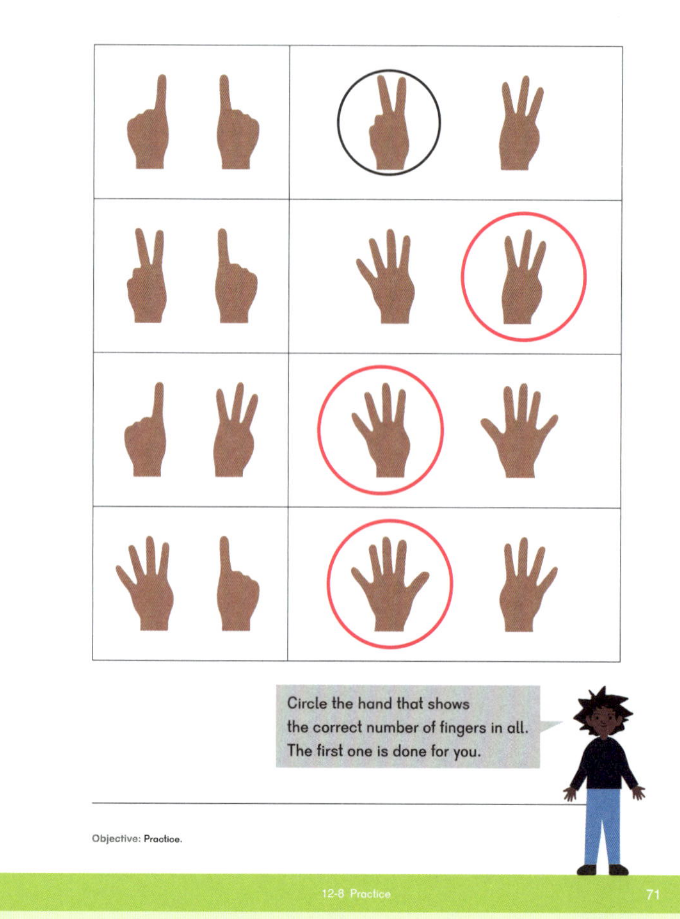

Exercise 8 • page 71

Extend Play

I Wish I Had 5 on My Card: Have students play in pairs as directed in the previous lesson.

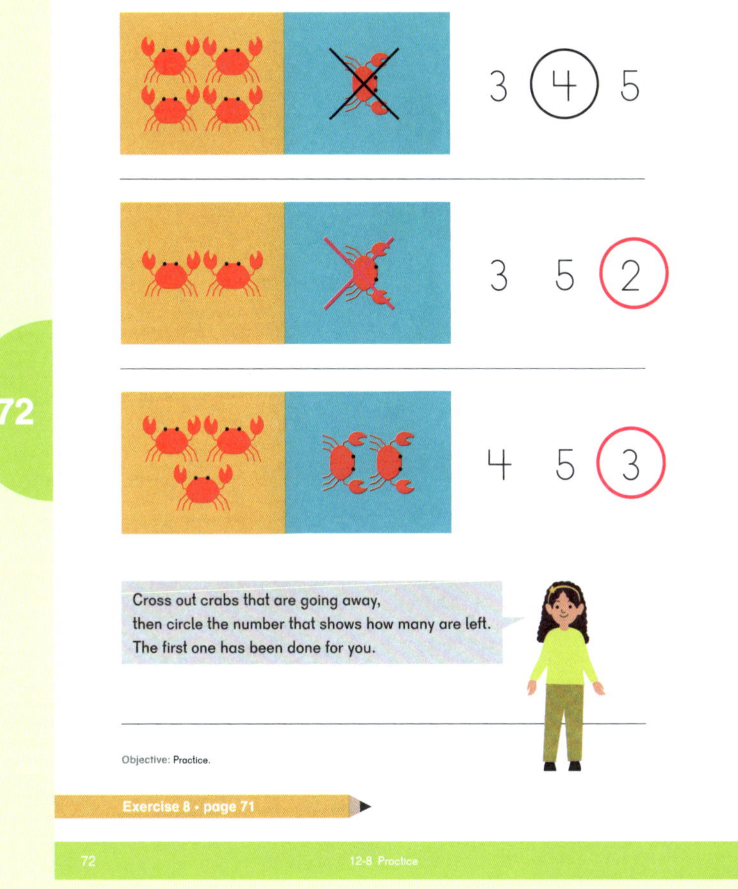

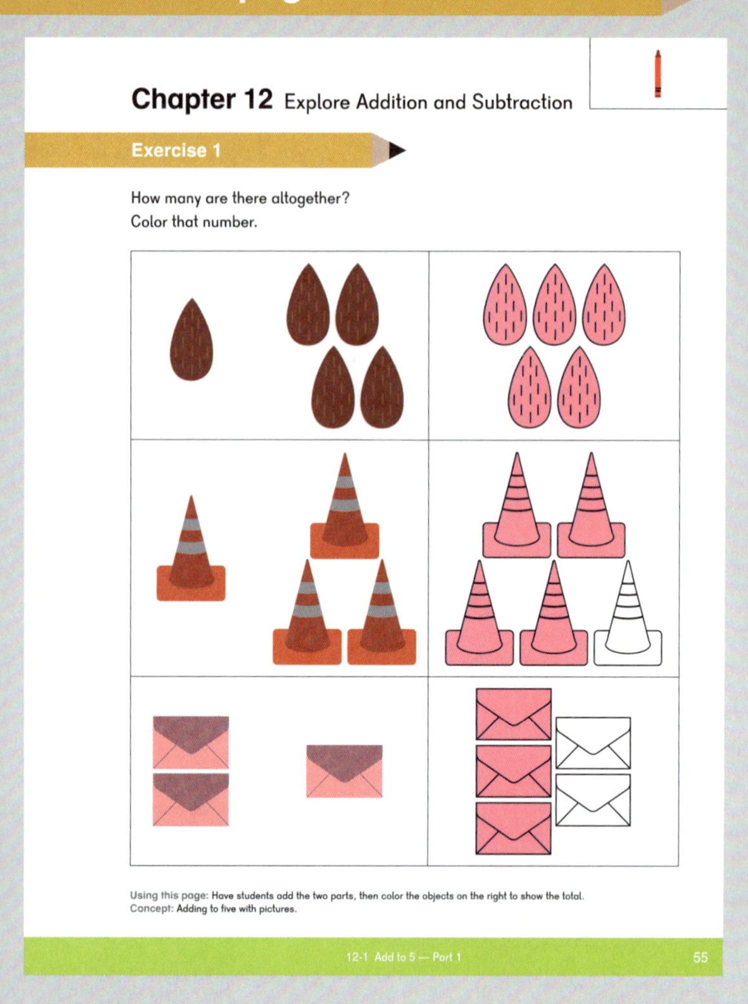
Chapter 12 Explore Addition and Subtraction
Exercise 1
How many are there altogether?
Color that number.
Using this page: Have students add the two parts, then color the objects on the right to show the total.
Concept: Adding to five with pictures.
12-1 Add to 5 — Part 1
55

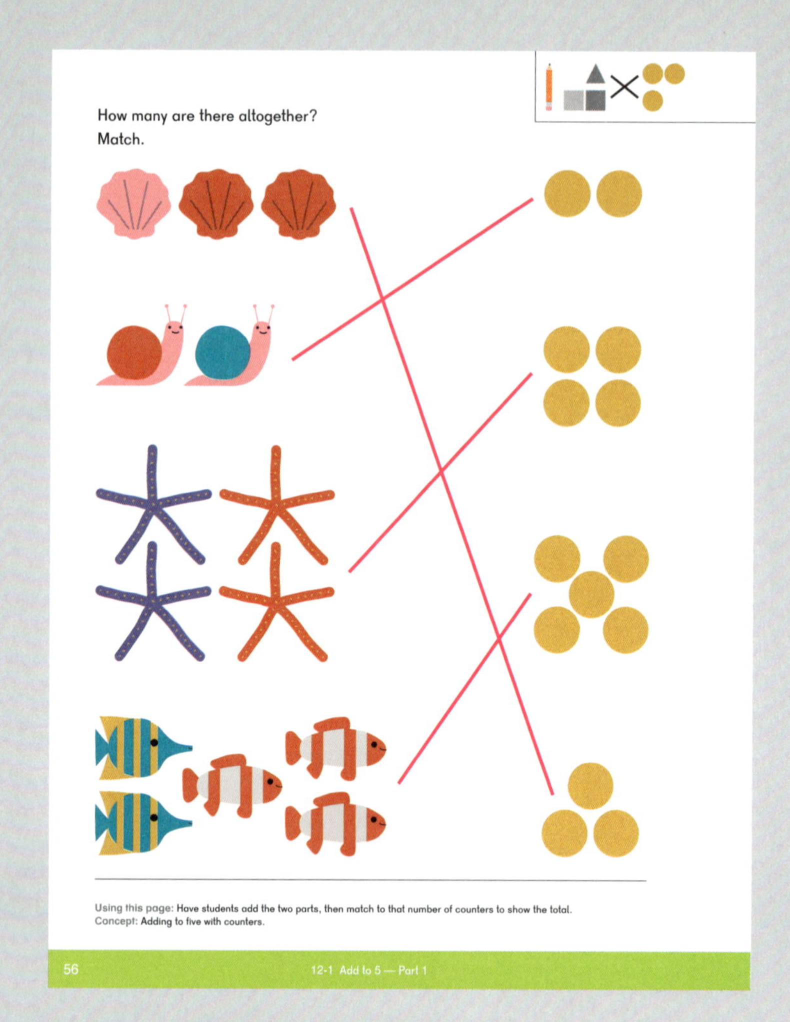
How many are there altogether?
Match.
Using this page: Have students add the two parts, then match to that number of counters to show the total.
Concept: Adding to five with counters.
56
12-1 Add to 5 — Part 1

Exercise 2 • pages 57–60

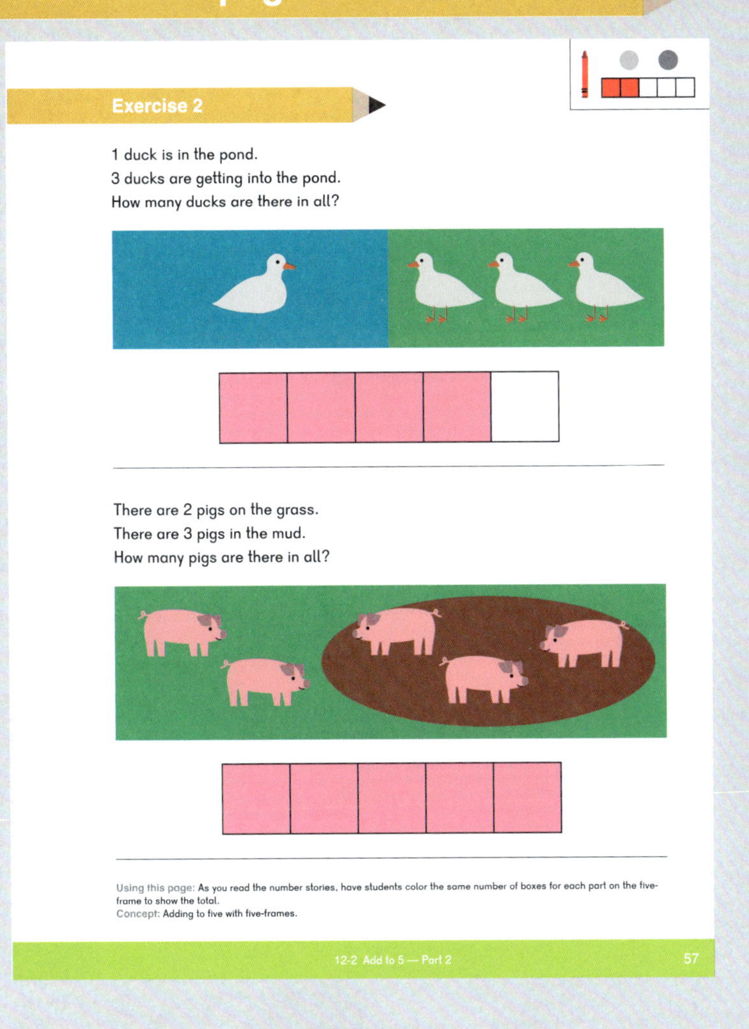

Exercise 2

1 duck is in the pond.
3 ducks are getting into the pond.
How many ducks are there in all?

There are 2 pigs on the grass.
There are 3 pigs in the mud.
How many pigs are there in all?

Using this page: As you read the number stories, have students color the same number of boxes for each part on the five-frame to show the total.
Concept: Adding to five with five-frames.

12-2 Add to 5 — Part 2 57

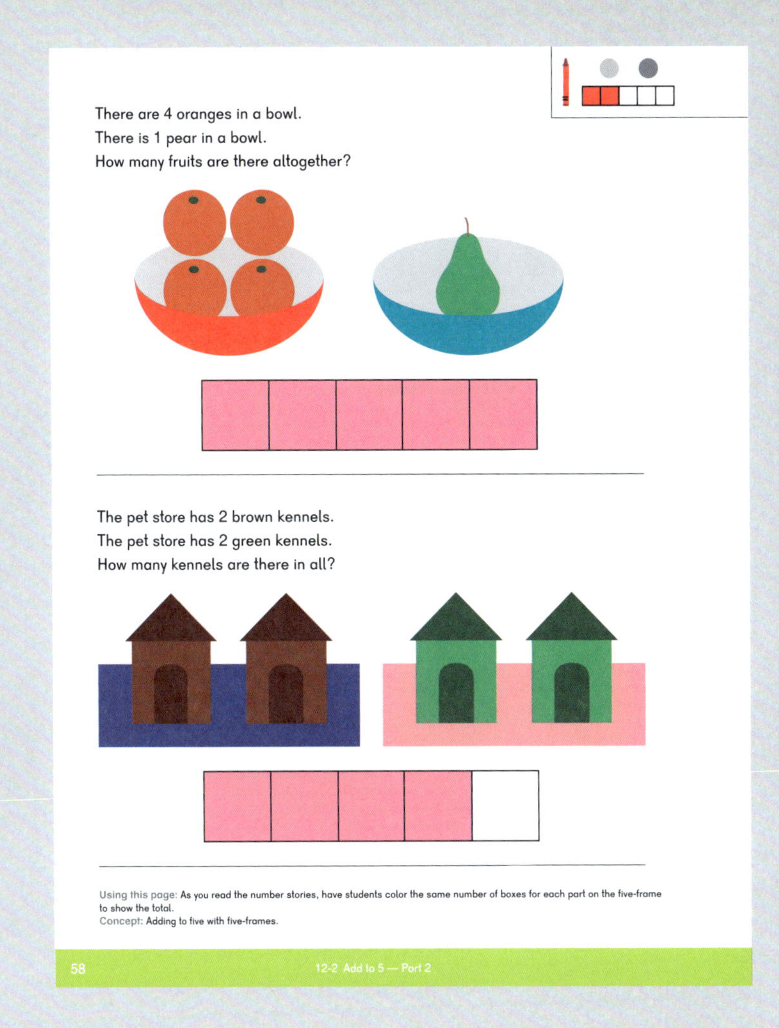

There are 4 oranges in a bowl.
There is 1 pear in a bowl.
How many fruits are there altogether?

The pet store has 2 brown kennels.
The pet store has 2 green kennels.
How many kennels are there in all?

Using this page: As you read the number stories, have students color the same number of boxes for each part on the five-frame to show the total.
Concept: Adding to five with five-frames.

58 12-2 Add to 5 — Part 2

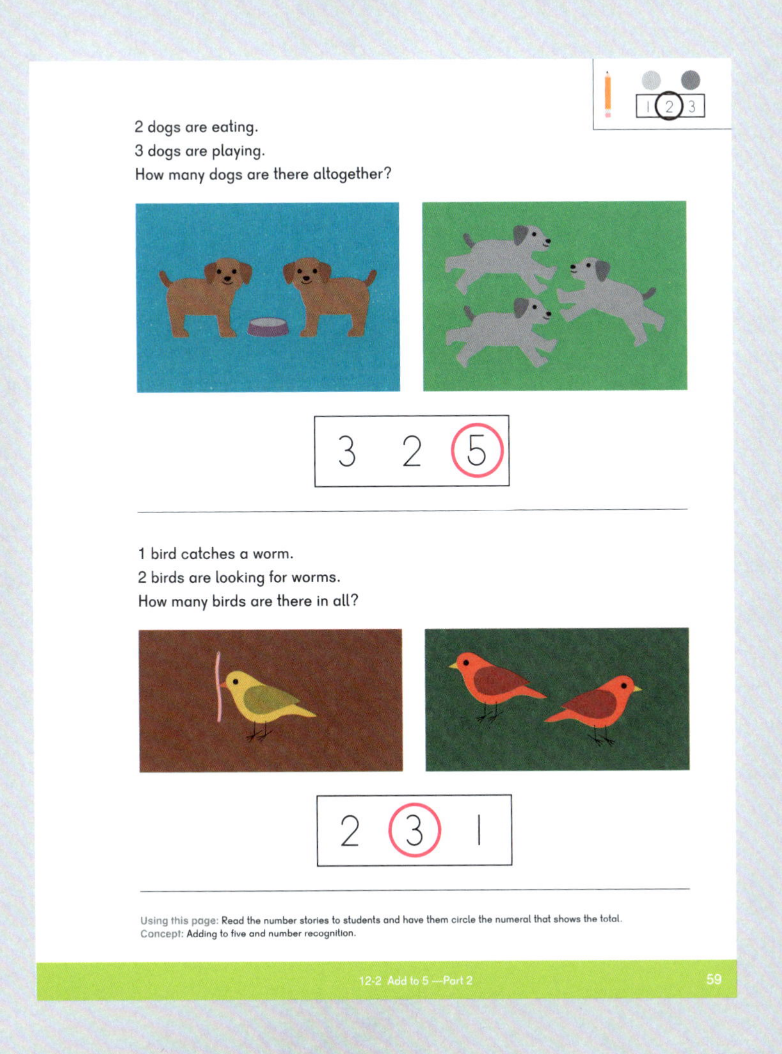

2 dogs are eating.
3 dogs are playing.
How many dogs are there altogether?

3 2 5

1 bird catches a worm.
2 birds are looking for worms.
How many birds are there in all?

2 3 1

Using this page: Read the number stories to students and have them circle the numeral that shows the total.
Concept: Adding to five and number recognition.

12-2 Add to 5 — Part 2 59

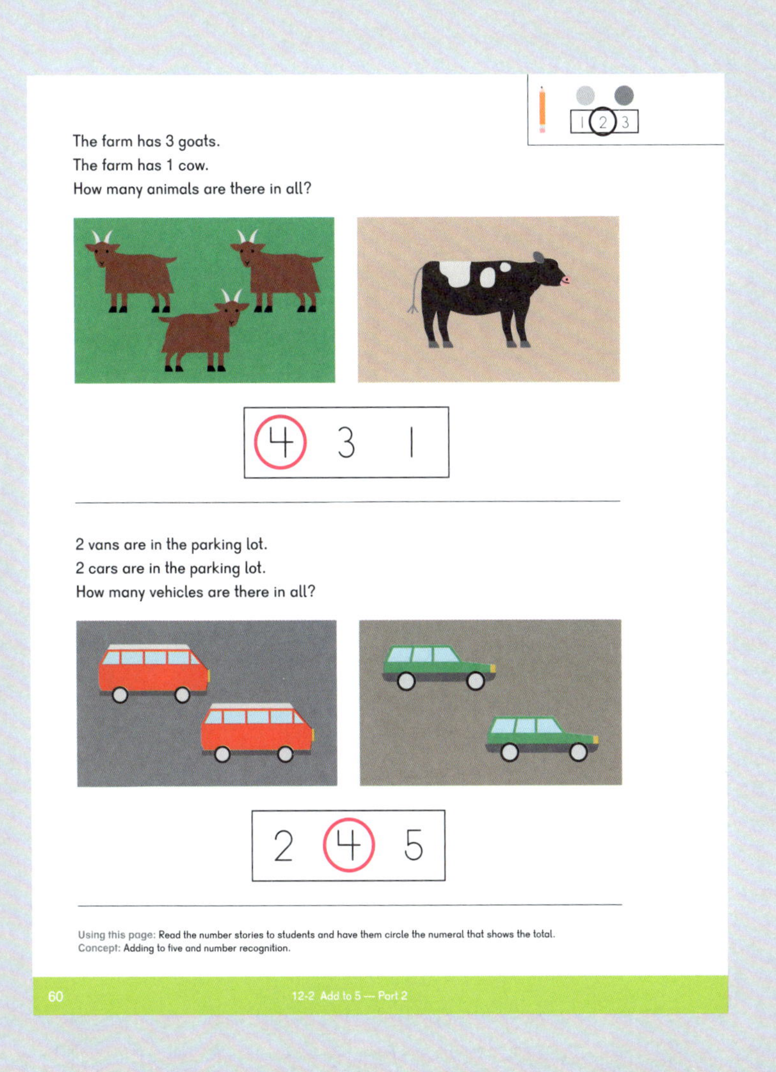

The farm has 3 goats.
The farm has 1 cow.
How many animals are there in all?

4 3 1

2 vans are in the parking lot.
2 cars are in the parking lot.
How many vehicles are there in all?

2 4 5

Using this page: Read the number stories to students and have them circle the numeral that shows the total.
Concept: Adding to five and number recognition.

60 12-2 Add to 5 — Part 2

Exercise 3 • pages 61–62

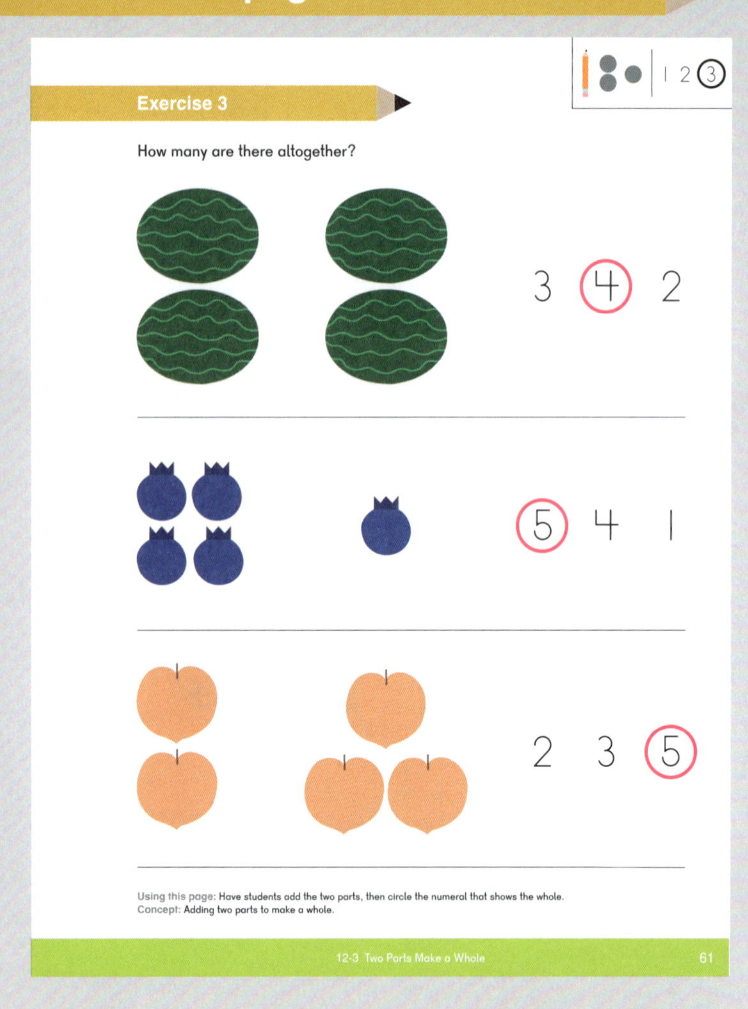

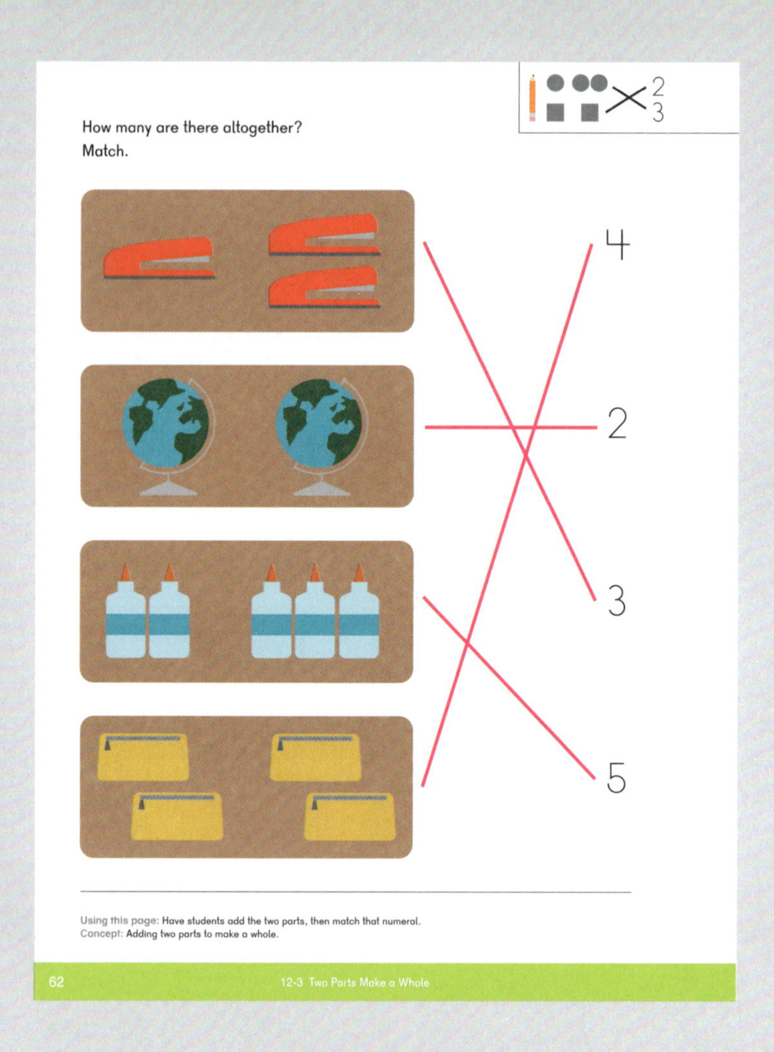

Exercise 4 • pages 63–64

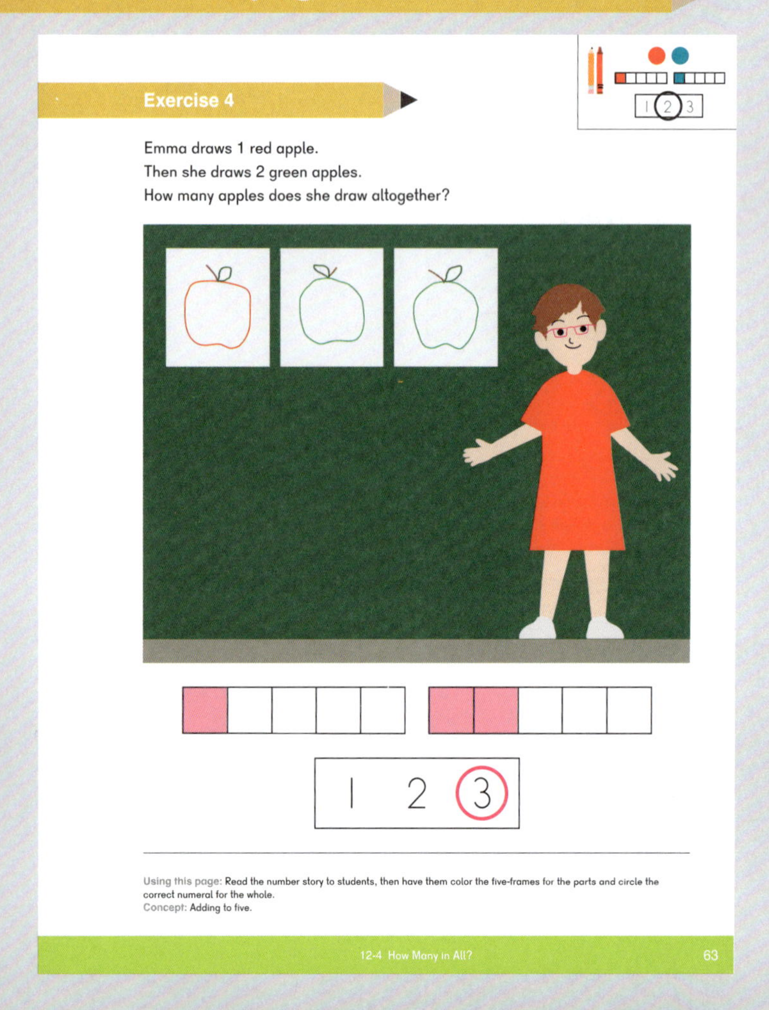

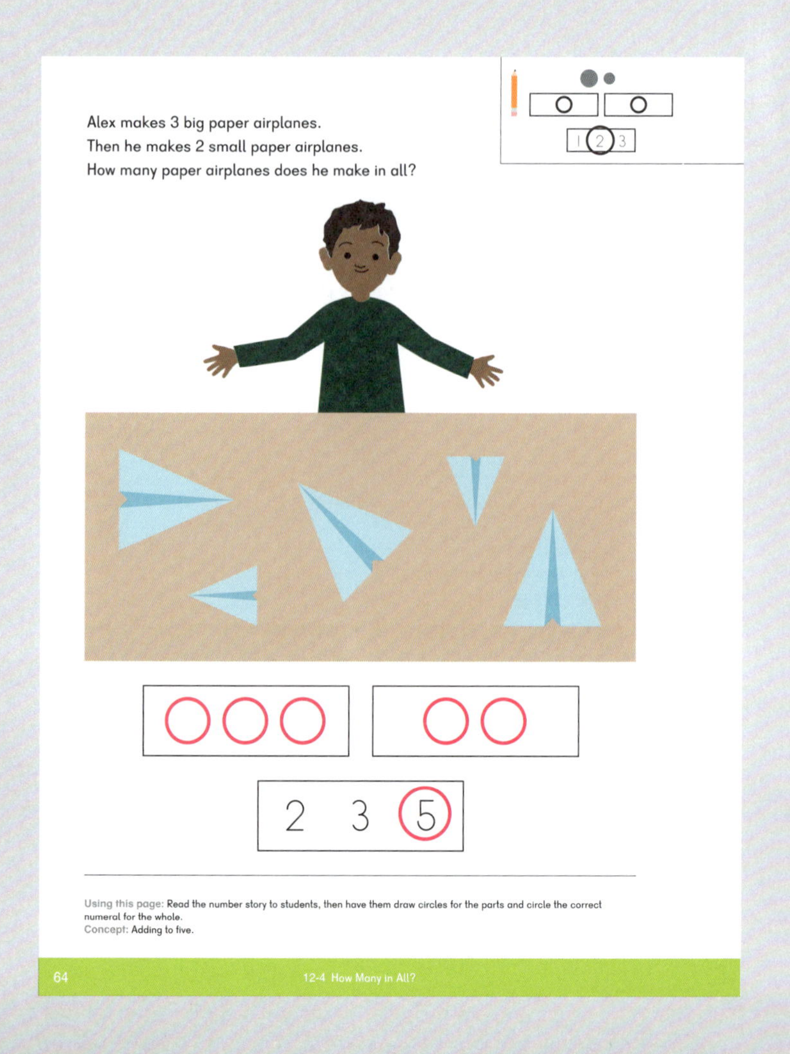

Exercise 5 • pages 65–66

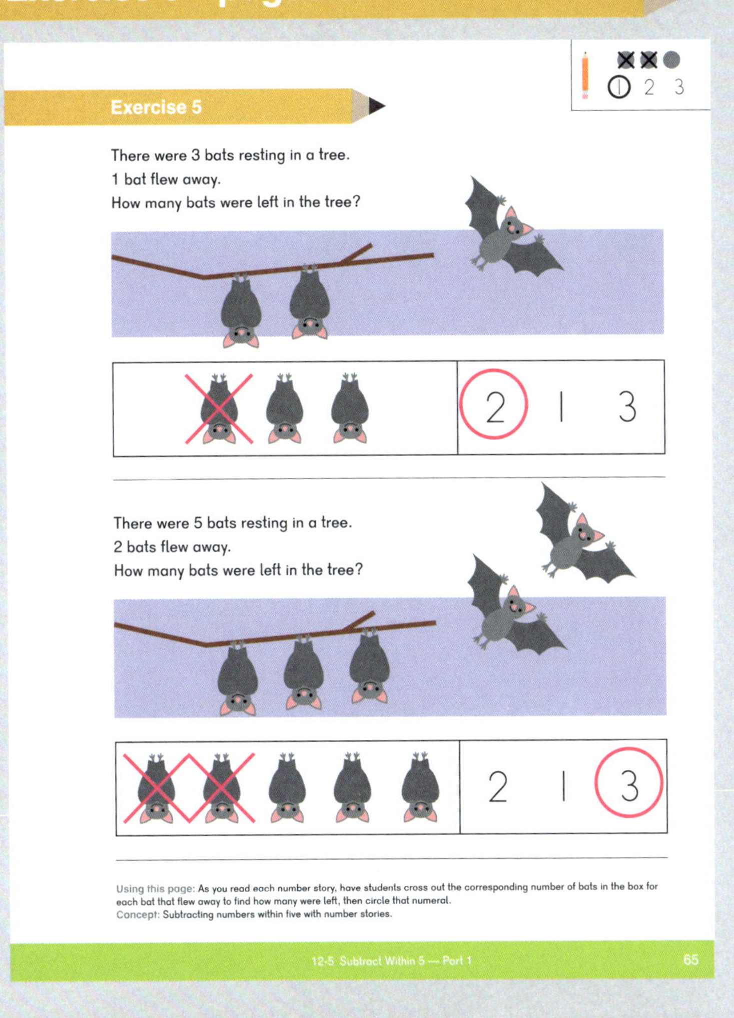

Exercise 5

There were 3 bats resting in a tree.
1 bat flew away.
How many bats were left in the tree?

2 1 3

There were 5 bats resting in a tree.
2 bats flew away.
How many bats were left in the tree?

2 1 3

Using this page: As you read each number story, have students cross out the corresponding number of bats in the box for each bat that flew away to find how many were left, then circle that numeral.
Concept: Subtracting numbers within five with number stories.

12-5 Subtract Within 5 — Part 1 65

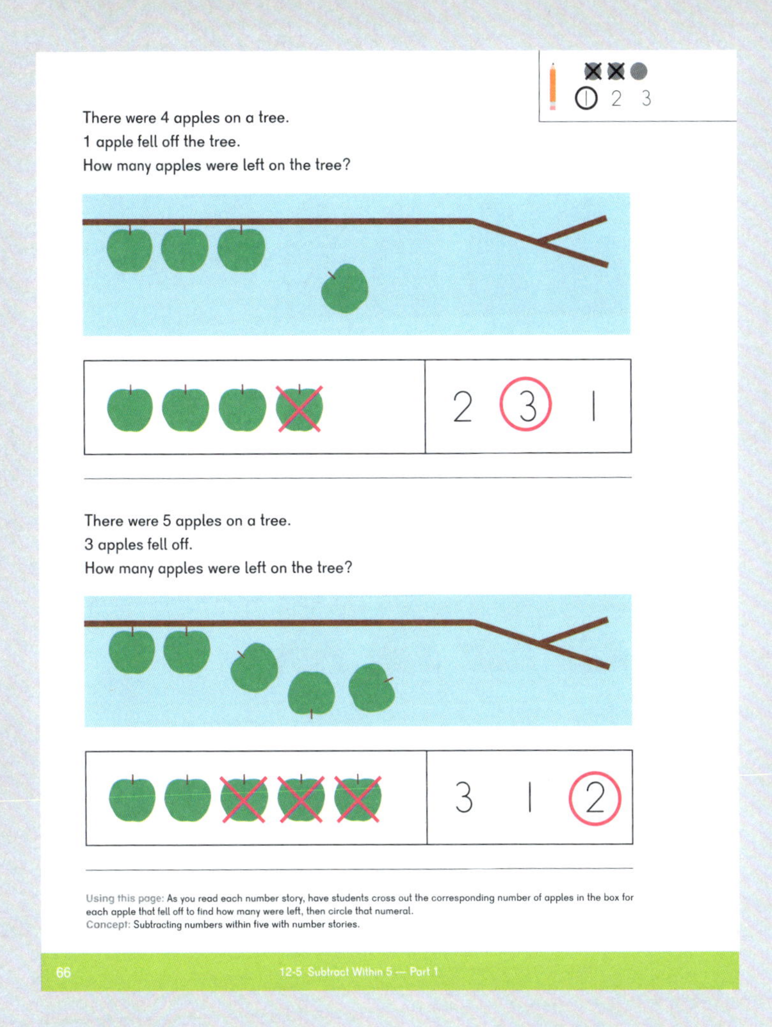

There were 4 apples on a tree.
1 apple fell off the tree.
How many apples were left on the tree?

2 3 1

There were 5 apples on a tree.
3 apples fell off.
How many apples were left on the tree?

3 1 2

Using this page: As you read each number story, have students cross out the corresponding number of apples in the box for each apple that fell off to find how many were left, then circle that numeral.
Concept: Subtracting numbers within five with number stories.

66 12-5 Subtract Within 5 — Part 1

Exercise 6 • pages 67–68

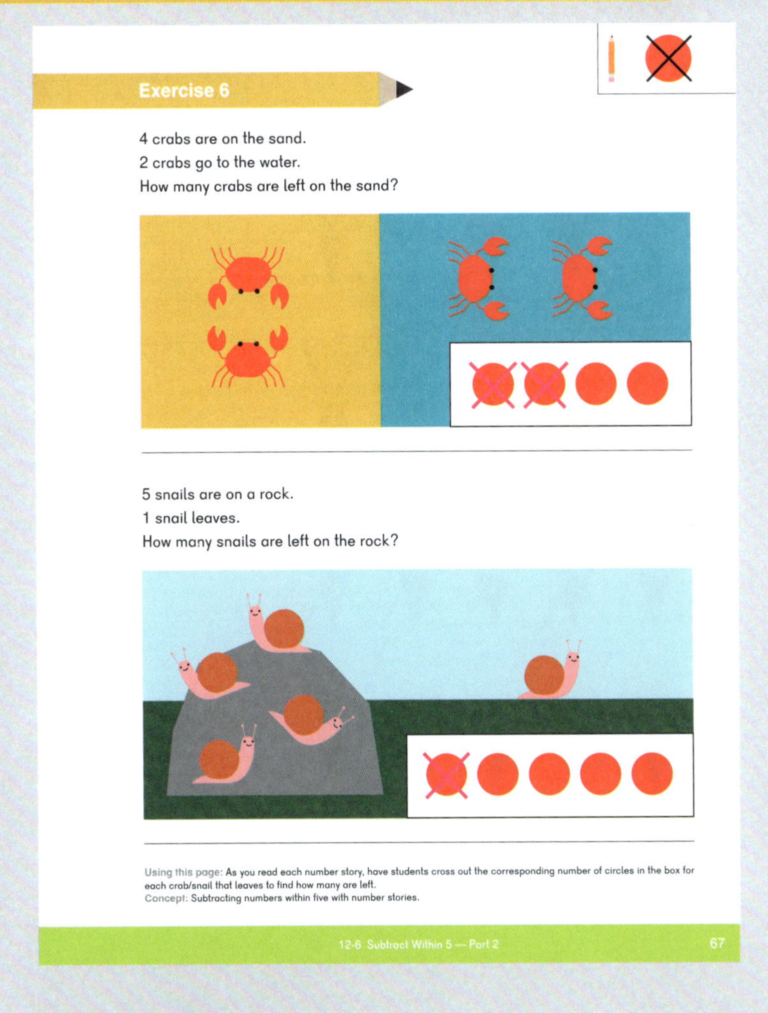

Exercise 6

4 crabs are on the sand.
2 crabs go to the water.
How many crabs are left on the sand?

5 snails are on a rock.
1 snail leaves.
How many snails are left on the rock?

Using this page: As you read each number story, have students cross out the corresponding number of circles in the box for each crab/snail that leaves to find how many are left.
Concept: Subtracting numbers within five with number stories.

12-6 Subtract Within 5 — Part 2 67

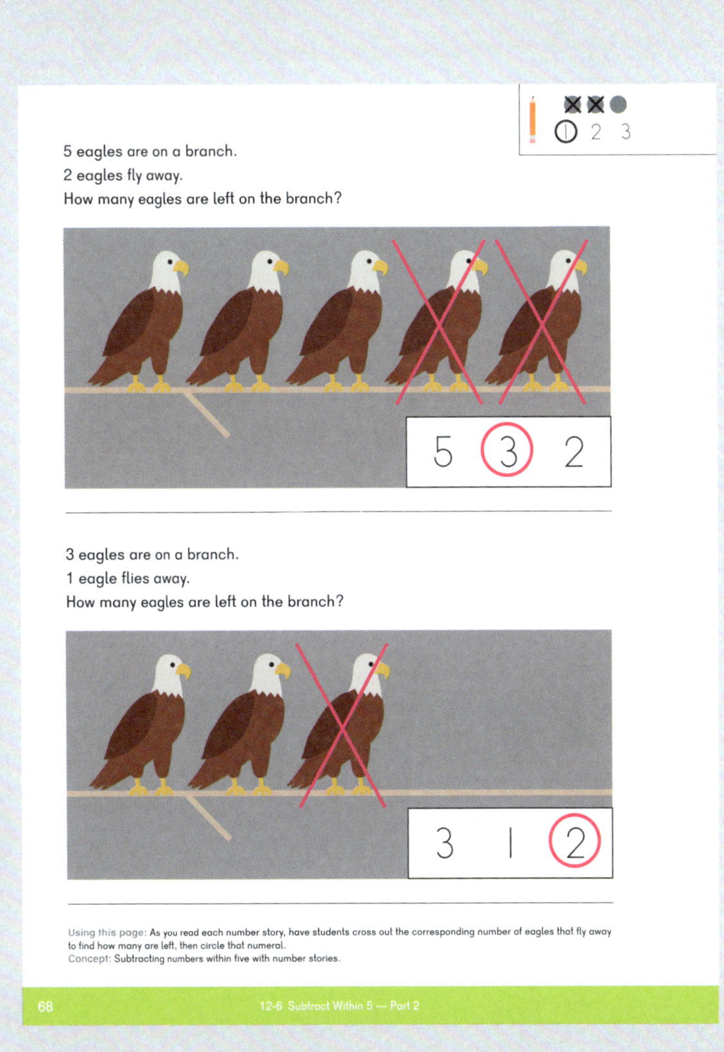

5 eagles are on a branch.
2 eagles fly away.
How many eagles are left on the branch?

5 3 2

3 eagles are on a branch.
1 eagle flies away.
How many eagles are left on the branch?

3 1 2

Using this page: As you read each number story, have students cross out the corresponding number of eagles that fly away to find how many are left, then circle that numeral.
Concept: Subtracting numbers within five with number stories.

68 12-6 Subtract Within 5 — Part 2

Exercise 7 • pages 69–70

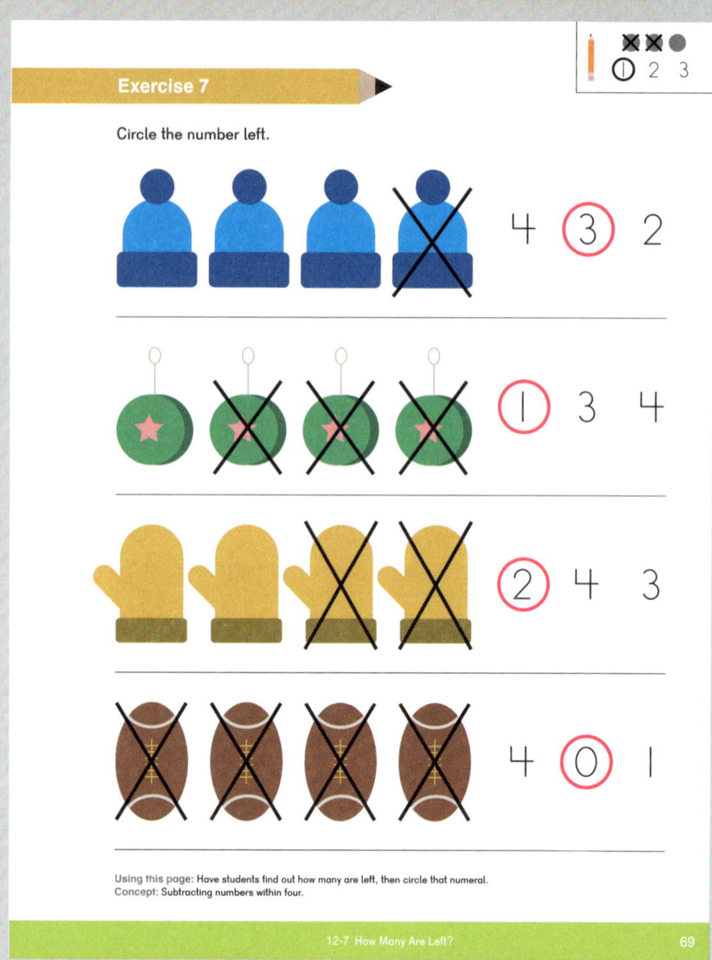

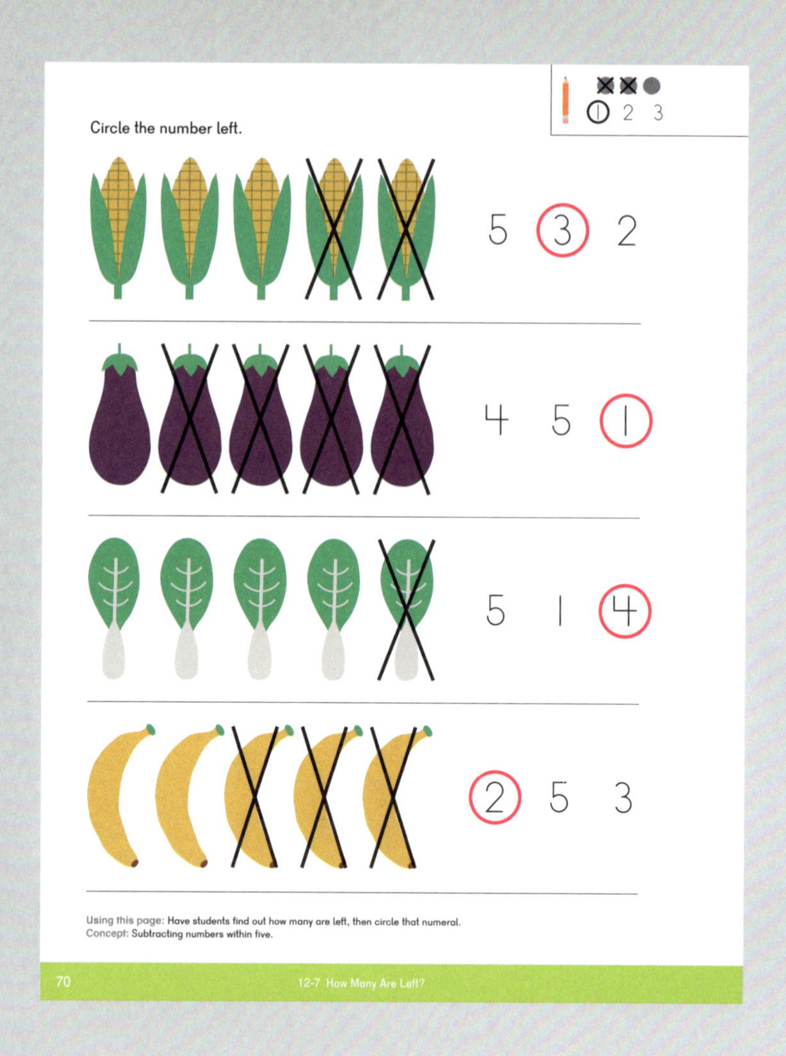

Exercise 8 • pages 71–72

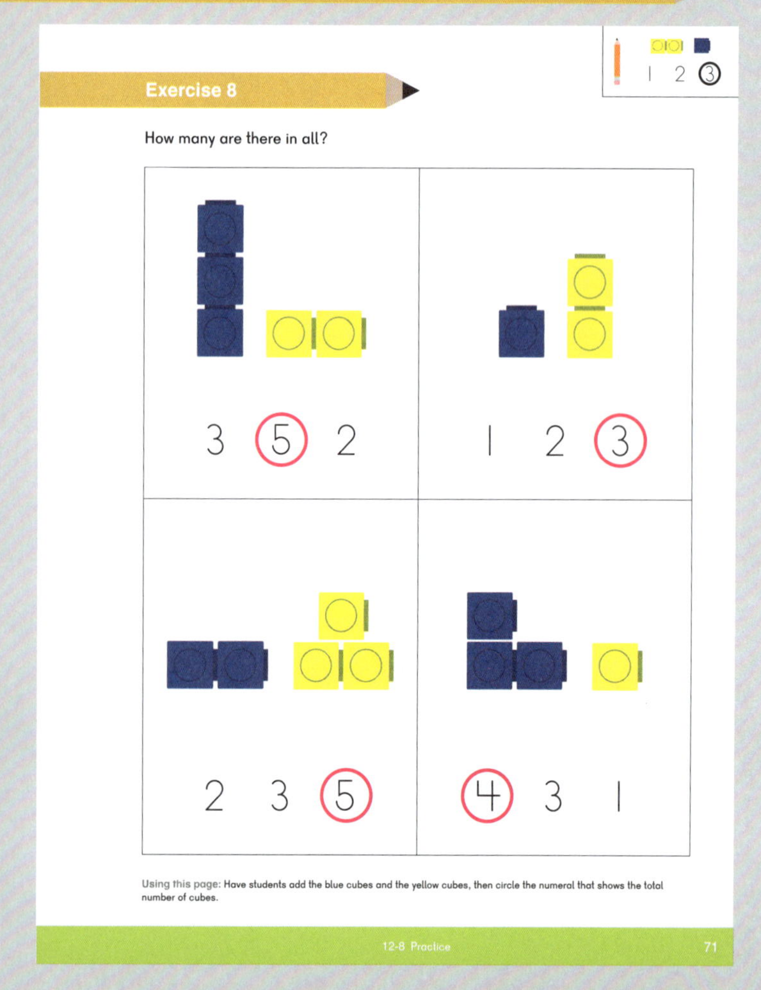

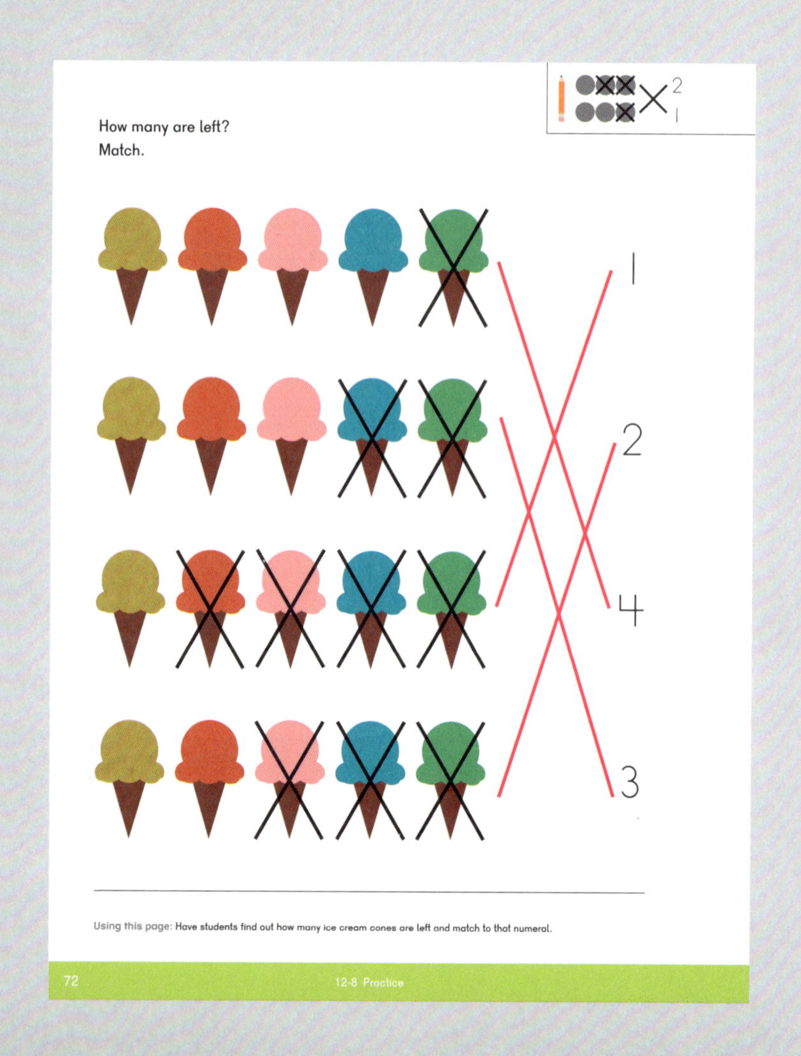

Suggested number of class periods: 23

	Lesson	Page	Resources	Objectives
	Chapter Opener	p. 119	TB: p. 73	
1	**Review 1** Match and Color	p. 120	TB: p. 74 WB: p. 73	Identify objects that are the same and not the same. Draw squares and circles. Recognize the colors blue, red, brown, green, orange, pink, and yellow.
2	**Review 2** Big and Small	p. 122	TB: p. 77 WB: p. 75	Compare objects by size and color.
3	**Review 3** Heavy and Light	p. 123	TB: p. 78 WB: p. 77	Compare by weight.
4	**Review 4** Count to 5	p. 124	TB: p. 79 WB: p. 79	Count sets of objects up to five with one-to-one correspondence. Identify squares and triangles.
5	**Review 5** Count 5 Objects	p. 126	TB: p. 81 WB: p. 81	Count 5 objects with one-to-one correspondence. Identify circles.
6	**Review 6** 0	p. 127	TB: p. 82 WB: p. 83	Understand that an empty set is a set of zero.
7	**Review 7** Count Beads	p. 128	TB: p. 83 WB: p. 85	Count with one-to-one correspondence.
8	**Review 8** Patterns	p. 130	TB: p. 85 WB: p. 87	Identify and extend patterns.
9	**Review 9** Length	p. 131	TB: p. 86 WB: p. 89	Compare objects by length.
10	**Review 10** How Many?	p. 132	TB: p. 87 WB: p. 91	Subitize dice patterns. Recognize quantities on ten-frame cards.
11	**Review 11** Ordinal Numbers	p. 134	TB: p. 89 WB: p. 93	Identify first through fifth from a starting point.
12	**Review 12** Solids and Shapes	p. 135	TB: p. 90 WB: p. 95	Recognize spheres, cubes, and cylinders. Make a graph from a picture. Recognize circles, squares, rectangles, and triangles.

	Lesson	Page	Resources	Objectives
13	**Review 13** Which Set Has More?	p. 138	TB: p. 94 WB: p. 97	Compare two sets of objects to find which has more.
14	**Review 14** Which Set Has Fewer?	p. 139	TB: p. 95 WB: p. 99	Compare two sets of objects to find which has fewer.
15	**Review 15** Put Together	p. 140	TB: p. 96 WB: p. 101	Identify the whole when two parts are given.
16	**Review 16** Subtraction	p. 142	TB: p. 98 WB: p. 103	Subtract numbers within 5.
17	**Looking Ahead 1** Sequencing — Part 1	p. 143	TB: p. 99	Use logic to determine the sequence of events.
18	**Looking Ahead 2** Sequencing — Part 2	p. 144	TB: p. 100	Use logic to determine the sequence of events.
19	**Looking Ahead 3** Categorizing	p. 146	TB: p. 102	Categorize objects by use.
20	**Looking Ahead 4** Addition	p. 147	TB: p. 103	Add two parts to find the whole.
21	**Looking Ahead 5** Subtraction	p. 149	TB: p. 105	Take away one part from the whole to find the other part.
22	**Looking Ahead 6** Getting Ready to Write Numerals	p. 151	TB: p. 107	Draw vertical, horizontal, curved, and diagonal lines.
23	**Looking Ahead 7** Reading and Math	p. 152	TB: p. 108	Find the number of letters in simple words.
	Workbook Solutions	p. 153		

Chapter Vocabulary

- Diagonal
- Horizontal
- Vertical

Students were introduced to cardinal numbers 1 through 10 in **Dimensions Math® PK A** Chapters 4 through 7. They were also introduced to ordinal numbers, which describe positions and have nothing to do with quantity.

Chapter 13 includes two main goals. The first, and more important, is a cumulative review of topics explored in earlier chapters.

Metacognition, or thinking about thinking, is a critical skill. In mathematics, the ability to make connections between and among various topics is very important. The cumulative review is designed to help students make these connections.

The **Chapter Opener** reminds students of some of the topics they have previously learned. Spend time asking students to recall the various topics and how they relate to each other. For example, how do "parts and wholes" relate to "counting?"

Because Pre-Kindergarten students are not yet writing explanations in a journal reflecting upon their learning, model for them how to do this by making notes on large chart paper based on their comments and posting the notes in the room. Draw pictures to illustrate the reflections whenever possible. Use questioning techniques to help them make connections. For example, "When we sort, what are some of the things we think about to make our groups?" Students may answer: color, size, length, weight, use, texture, etc. Your notes on these reflections could show the word "Sort" in the middle of the page with spokes going to each of the ideas, and pictures to represent the ideas.

The second goal is to help prepare those students who are ready for upcoming topics in Kindergarten. The lessons are brief introductions to the topics. The content of the lessons is not expected to be mastered in Kindergarten. **Looking Ahead 4: Addition** and **Looking Ahead 5: Subtraction**, are included only for those students who are ready for the symbolic representation of equations (number sentences). **Dimensions Math® Kindergarten B** Chapters 9, 10, and 11 contain thorough explorations of these topics.

Key Points

Because making connections is so important, continue to emphasize discussion. Many lessons in this chapter suggest the use of a recording device. If you choose to follow this suggestion, listen to students' recordings. Look for clear explanations and number stories that are understandable and relate to the requested topic or operation. Later, have students listen to their own recordings and discuss your comments on each.

Review 13: Which Set Has More and **Review 14: Which Set Has Fewer** both have students playing **Musical Chairs** for **Whole Group Play**. This is intentional as it will help students understand the relationship between "more" and "fewer."

Set up small groups of chairs rather than having all students playing in one large group so that there is less waiting time for students who get "out."

Because Chapter 13 is covered at the end of the school year, and to help students connect topics learned earlier, many center ideas are repeated from earlier chapters. **Looking Ahead 1, 2,** and **3** suggest **Free Center Days**. Begin interviewing students regarding their favorite math centers. You may decide to do a graph of them so students see which centers are most popular. On those three days, set out 5 or 6 of the all-time favorites and let students choose which center they want to visit.

Looking Ahead 1: Sequencing — Part 1 involves growing 3 plants over the course of this chapter. Students' observations of the plant growth will help them to assimilate the topic. Plant some radish (or other fast-growing) seeds in a see-through plastic cup with your students at the beginning of this Chapter. Repeat on day 5 or 6. Repeat on day 10 or 12. Observing the plant growth will help students with the lesson.

Materials

- Agility cones
- Aluminum foil
- Animal counters
- Balance scales
- Building materials
- Capes
- Cardboard tubes
- Cards with students' names on them
- Chairs
- Clothespins
- Container
- Cotton balls
- Counters
- Crayons or markers, including blue, red, brown, green, orange, pink, and yellow
- Cubes, cylinders, and spheres from geometry sets and manipulative kits
- Daubers of different colors
- Dice
- Dominoes
- Drinking straws of two colors
- Drums
- Dry rice, dried beans, and/or sand
- Embroidery hoops
- Equilateral, isosceles, and scalene triangles cut out of paper
- Five-divot and ten-divot egg cartons
- Fly swatters
- Geoboards and rubber bands
- Growing plants at different stages
- Hula hoops
- Jump ropes of different lengths
- Lacing cards (square, triangle, and numbers 0 to 10)
- Large, dull-tip needles
- Lego® or DUPLO® style interlocking blocks
- Linking cubes
- Magazines to cut up
- Modified dice with sides labeled: +1, +1, +1, +2, +2, +2
- Musical instruments
- Numerous everyday objects that come in pairs
- Objects that are the same except for size and/or color
- Paper cups
- Paper plates
- Paper shapes: yellow circle, green square, orange rectangle, and purple triangle
- Paper strips of different lengths
- Part-whole mats
- Pieces of yarn or string of different lengths
- Pipe cleaners
- Plastic bags
- Plastic eggs of different colors
- Plastic needles
- Play dough
- Pool noodles cut into coins
- Real-life examples of the cubes, rectangular prisms, cylinders, and spheres
- Rough, smooth, soft, and hard objects to create crayon rubbings
- Small objects including those with varying textures, sizes, colors/patterns, and relationships
- Small objects of obviously different weights, at least 1 per pair of students
- Small pieces of fabric
- Sorting mats
- Square pieces of white paper
- Square tile or square paper cutouts of different colors
- Squares of large-weave burlap
- Stickers of several types
- Strawberries
- Strips of brown paper
- Stuffed toys or bean bags
- Toy bats (animals)
- Tubs of sand (or access to sandbox)
- Two-color counters
- Xylophones
- Yarn

Note: Materials for Activities will be listed in detail in each lesson.

Blackline Masters

- 0 Art Paper
- Addition and Equal Symbol Cards
- Addition Facts Cards
- Blank Five-frame
- Circle Match Cards
- Domino Cards
- Dot Cards
- Number Cards
- Number Cards — Large
- Paper Cutouts — Review 4
- Parts and Whole Mats
- Picture Cards
- Square-Triangle Flash Cards
- Subtraction and Equal Symbol Cards
- Ten-frame Cards
- Word Cards
- Word-Picture Cards

Storybooks

- *Stellaluna* by Janell Cannon
- *Big Shark, Little Shark* by Anna Membrino
- *Big Dog ... Little Dog* by P.D. Eastman
- *Go, Dog. Go!* by P.D. Eastman
- *Thumbelina* by Hans Christian Andersen
- *Shapes, Shapes, Shapes* by Tana Hoban
- *Goldilocks and the Three Bears*
- *Snow White* by The Brothers Grimm
- *The Very Hungry Caterpillar* by Eric Carle
- *Henry the Fourth* by Stuart J. Murphy
- *Pete the Cat and His Four Groovy Buttons* by James Dean

Optional Snacks

- "Turtles" made of kiwi slices (body) and green grape halves (head and legs)
- Apple slices
- Banana or kiwi slices
- Carrot sticks
- Celery sticks
- Cheese sticks
- Crackers of different sizes
- Fish crackers
- Food in the shapes of cubes, cylinders, and spheres, such as cheese cubes, jello cubes, marshmallows, grapes, and puff cereal
- Fruits of different colors
- Fruits that bats like, such as mangoes, bananas, and avocados
- Grapes, red and green
- Hard-boiled eggs
- Mini cheese balls
- Nut-free trail mix
- Orange slices
- Pretzel sticks and rods
- Raisins
- Round crackers
- Square crackers
- Square graham crackers covered with cream cheese and raisins in dice patterns
- Strawberries
- Triangle and square crackers

Letters Home

- Chapter 13 Letter

The song “Mary Had a Little Lamb” is referenced on page 134 of this Teacher’s Guide in **Review 11**. The lesson is a review of ordinal numbers. The instructions for students do not teach keyboard notes, but are as follows:

Third second first second third third third
second second second third fifth fifth
Third second first second third third third
third second second third second first

“First” is the thumb on the right hand. “Fifth” is the pinkie on the right hand.

The actual notes are as follows:

E, D, C, D, E, E, E
D, D, D, E, G, G
E, D, C, D, E, E, E
E, D, D, E, D, C

Mary Had a Little Lamb

Mary had a little lamb,
It’s fleece was white as snow;
And everywhere that Mary went
The lamb was sure to go.

He followed her to school one day
Which was against the rule;
It made the children laugh and play,
To see a lamb at school.

And so the teacher turned him out,
But still he lingered near;
And waited patiently about
Till Mary did appear.

“What makes the lamb love Mary so?”
The eager children cry;
“Why, Mary loves the lamb, you know,”
The teacher did reply.

The song “Five Little Bats” is included in **Review 16** on page 143 of this Teacher’s Guide.

Five Little Bats

Five little bats came flying in the door,
One flew away and that left four.

Four little bats hiding in a tree,
One flew away and that left three.

Three little bats looking down at you,
One flew away and that left two.

Two little bats hiding from the sun,
One flew away and that left one.

One little bat hanging all alone,
He flew away and then there were none.

Lesson Materials

- Five-divot and ten-divot egg cartons
- Sorting mats
- Part-whole mats
- Small objects including those with varying textures, sizes, colors/patterns, and relationships
- Optional snack: fruits of different colors

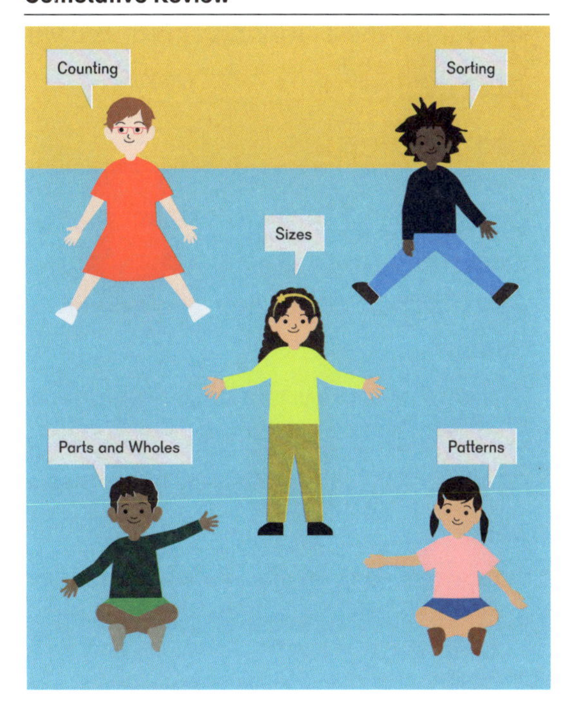

Explore

Place numerous small objects on the floor along with five-divot and ten-divot egg cartons, sorting mats, and part-whole mats. Some of the objects should be the same, some should be similar, and some should be unique, but set out no more than 5 of any one item. Allow students to explore the objects as they wish.

Learn

Ask students to explain what they did with the objects.

Have students look at page 73. Read each character's speech bubble, one at a time, and ask students what they recall about each topic. Then ask them how the topics relate. You may have to suggest connections to get them started. For examples, objects can be sorted, then counted, then arranged in patterns.

Extend Learn

What Do You Remember?: Have students choose one of the topics named on page 73. Then have them use the objects from **Explore** to demonstrate what they remember about the topic.

Objectives

- Identify objects that are the same and not the same.
- Draw squares and circles.
- Recognize the colors blue, red, brown, green, orange, pink, and yellow.

Lesson Materials

- Containers full of small, similar objects in blue, red, brown, green, orange, pink, and yellow to sort, 1 per pair of students
- Container
- Crayons or markers, including blue, red, brown, green, orange, pink, and yellow
- Paint in colors listed above, plus black and white
- Optional snack: fruits of different colors

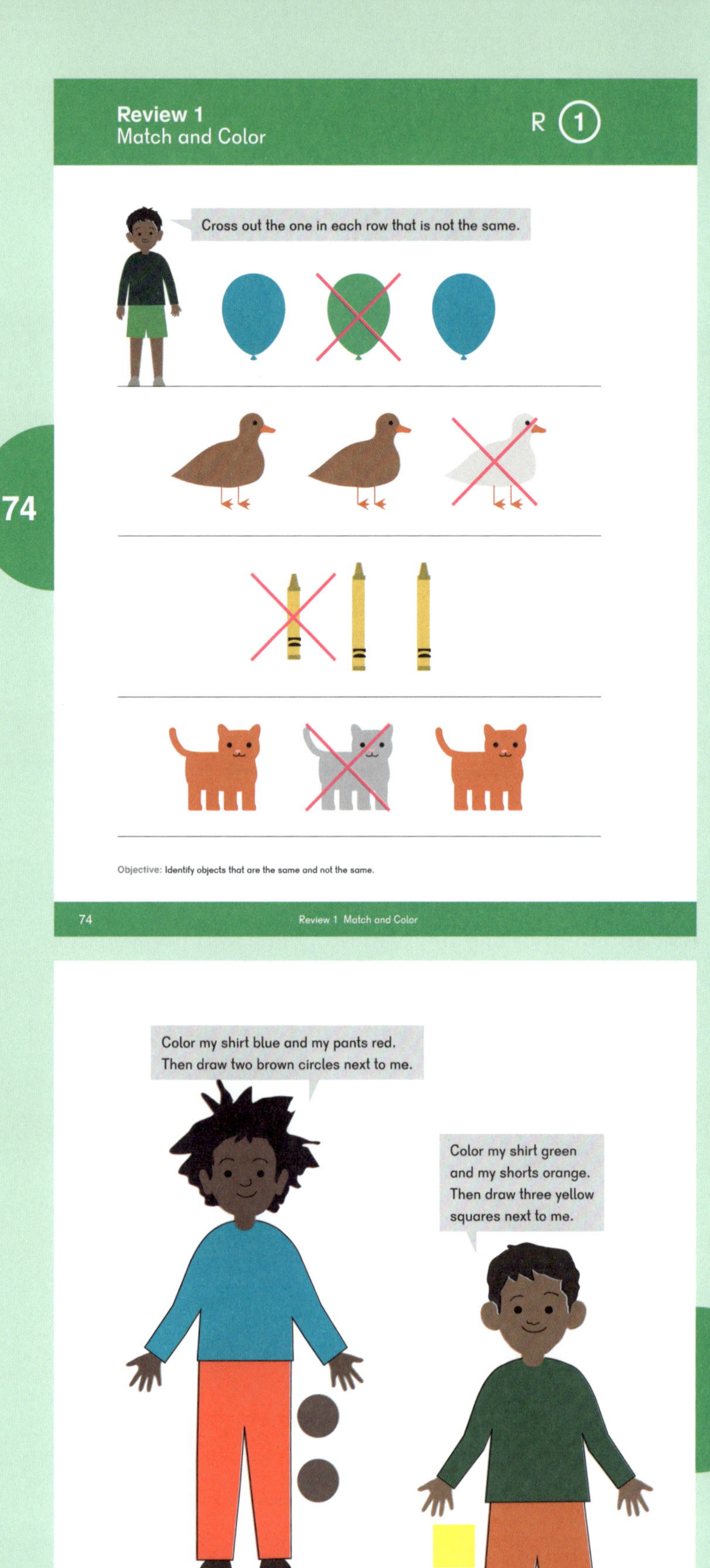

Explore

Give each pair of students a container of objects to sort into groups. Have them explain the reason for their sorts, using the terms "same" and "not the same." Have them sort the objects in different ways, each time explaining the reason for the sorts.

Learn

Hold up either crayons or markers, including blue, red, brown, green, orange, pink, and yellow, and have students identify each color. As you hold up each characters' favorite colors, ask students if they remember which character especially likes each of those colors. (Emma — red, Dion — blue, Sofia — yellow, and Alex — green).

Have students look at page 74 and identify which object in each row is not the same as the others and tell why. Read Alex's speech bubble and have them complete the task.

Draw a square in the air and have students do the same. Repeat with a circle.

Have students look at page 75. Read each character's speech bubble to students and have them complete the task.

Have students look at page 76. Review the color key with them. Ask them what color they will use to color the sun. If you are not sure that they understand the directions, repeat the question for the other objects on the page. Read Alex's speech bubble and have students complete the task.

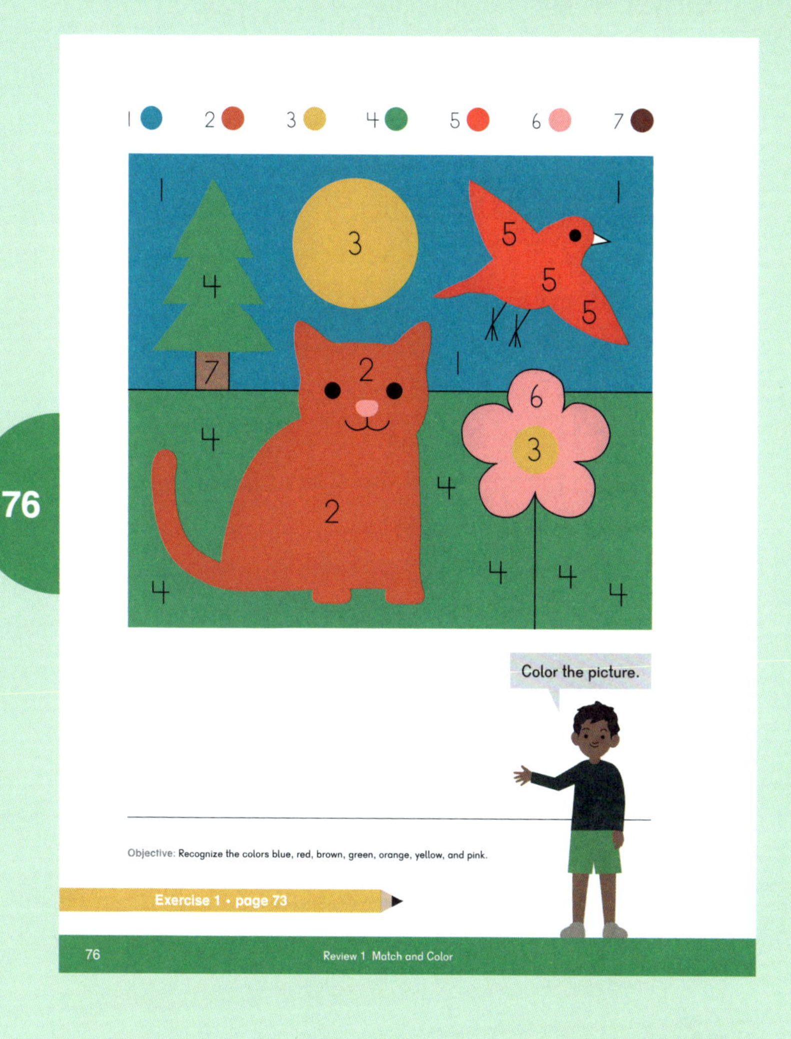

Whole Group Play

I Spy: Play "I Spy" using the colors reviewed as part of the clues. Then, have students sort objects by color.

Small Group Center Play

Color Shades: First, have students choose a color. Then have them create different shades of the color by adding small amounts of white and black paint. They can use the various shades of one color to produce a masterpiece.

Dress-Up: Have students choose a character and dress up in clothing of that character's favorite color.

Exercise 1 • page 73

Extend Learn

Can You Make What I Made?: Use crayons in the colors of this lesson to create a picture using squares and circles. Describe your picture and see if students can create something similar.

Objective

- Compare objects by size and color.

Lesson Materials

- Containers to sort into
- Objects that are the same except for size and/or color
- Storybook about sizes (See suggestions on page 118 of this Teacher's Guide)
- Lego® or DUPLO® style interlocking blocks
- Rough, smooth, soft, and hard objects to create crayon rubbings
- Optional snack: crackers of different sizes

77

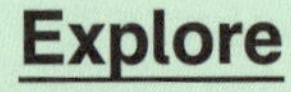

Explore

Give each pair of students a container of objects to sort into groups by size. Have them explain the reason for their sorts, using the terms "big" and "small." Have them sort the objects again by color.

Learn

Read a storybook about sizes. Have students act out the story.

Have students look at page 77 and describe each cup on the page by color and size. Read Sofia's speech bubble and have them complete the task.

Whole Group Play

Size and Color Simon Says: Play Simon Says using size and/or color as part of the directions. For example, "Simon says, stand next to something that is small and blue." Then, have students sort objects by size.

Small Group Center Play

Building: Build a big structure with Lego® or DUPLO® style interlocking blocks. Have students tell which structure is big and which is small.

Tell students that your structure is big and that they are to build a small one.

Paint: Provide big and small pieces of paper and paint in the colors reviewed for students.

Texture Rub Art: Have students use a crayon to rub one big and one small object.

Exercise 2 • page 75

Extend Learn

I'm Thirsty: Have students decide which type of drinks would work best in the containers shown on page 77 and explain their reasoning.

DUPLO® and Lego® are trademarks of the Lego Group of companies.

Objective

- Compare by weight.

Lesson Materials

- Small objects of obviously different weights, at least 1 per pair of students
- Building materials of different weights
- Balance scales
- Cotton balls
- Paper plates
- Clothespins
- Linking cubes
- Optional snack: bananas or grapes for students to weigh before eating

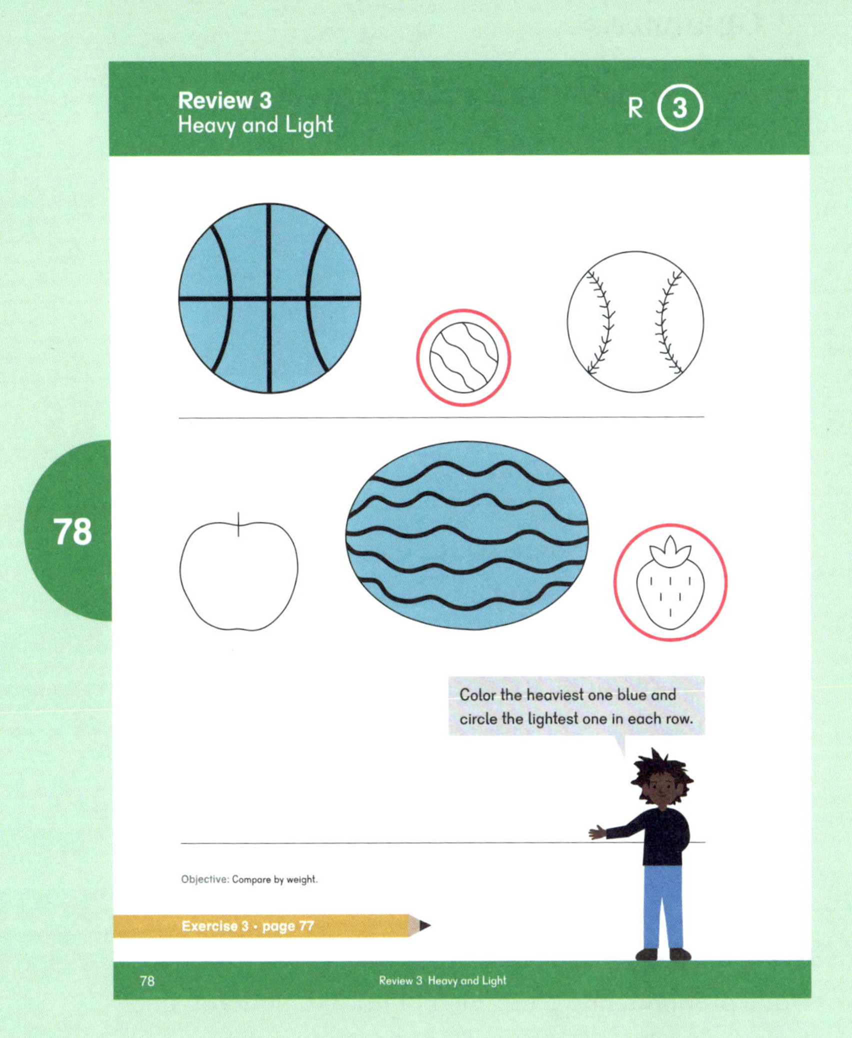

Explore

Set out objects to weigh. Have students compare the weights of the objects by holding them. Allow them to use the balance scales to weigh various objects.

Learn

Have pairs of students weigh four objects on balance scales and order them from lightest to heaviest.

Have students look at page 78 and identify the objects in each row. Read Dion's speech bubble to them and have them complete the task.

Whole Group Play

Give each pair of students a classroom object and tell them their task is to arrange all of the objects in order from lightest to heaviest. Students must decide where objects belong and possibly move objects from one place to another. Some objects may have to be weighed on balance scales.

Small Group Center Play

Cotton Ball Painting: Set out cotton balls, paints, clothespins, and paper plates. Show students how to use a clothespin to pick up a cotton ball. Dip the cotton ball in paint and transfer the paint to the paper plate. Have students create masterpieces.

Heavy Materials: Have students build the tallest structure they can and explain why they chose to put their building materials where they did. Should heavier objects go on the top or the bottom? Why?

Exercise 3 • page 77

Extend Play

Make it Balance: Give each pair of students linking cubes and several small toys from the classroom. Using balance scales, have them figure out how many linking cubes weigh the same as each toy.

Objectives

- Count sets of objects up to five with one-to-one correspondence.
- Identify squares and triangles.

Lesson Materials

- Paper Cutouts — Review 4 (BLM)
- Square-Triangle Flash Cards (BLM)
- Lacing cards (square and triangle)
- Plastic needles
- Yarn
- Geoboards and rubber bands
- Magazines to cut up
- Equilateral, isosceles, and scalene triangles cut out of paper
- Optional snack: triangle and square crackers

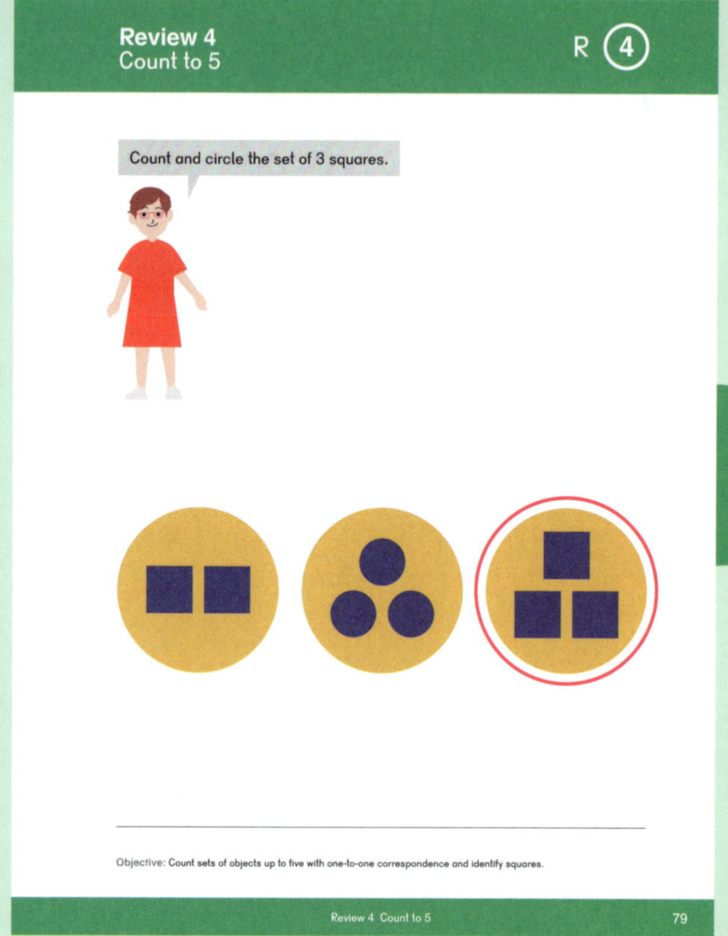

79

Explore

Have students find examples of up to 5 squares and triangles in the classroom and bring them to group. Have them tell how they know which shape is which and how many of each they found.

Learn

Sing "Circle, Square, Rectangle, Triangle" (VR) with students. Use your finger in the air to draw each shape as you sing about it and have students do the same.

Have students look at page 79. Read Emma's speech bubble and have students complete the task.

Have students look at page 80. Read Alex's speech bubble and have students complete the task. On both pages, have students name the shapes and the number of shapes in the groups they do not circle.

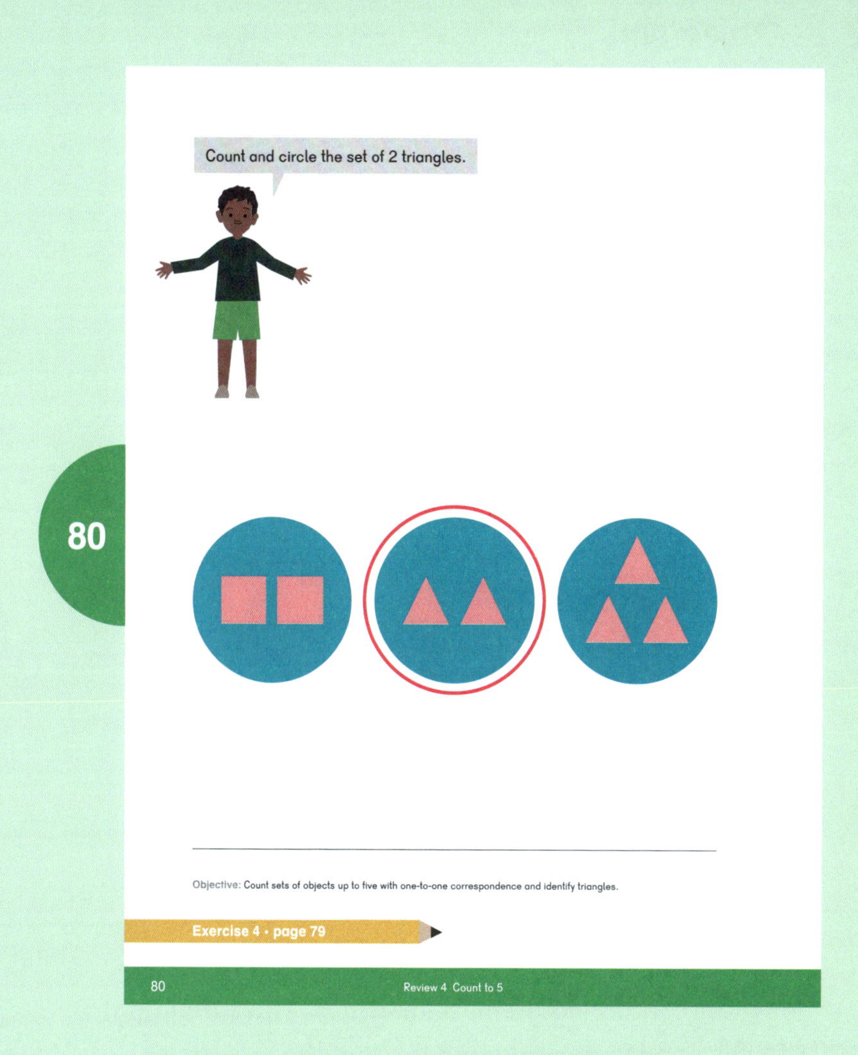

Whole Group Play

Flash It!: Show Square-Triangle Flash Cards (BLM) to students and have them call out the shape on the card (square or triangle) and the number of shapes on the card.

Small Group Center Play

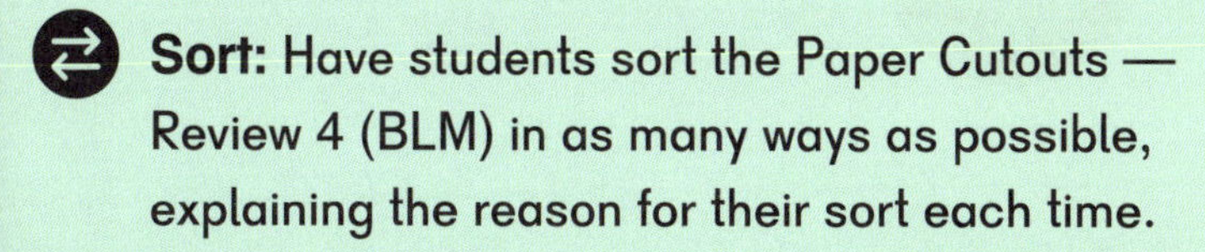

Sort: Have students sort the Paper Cutouts — Review 4 (BLM) in as many ways as possible, explaining the reason for their sort each time.

Shape Collages: Provide magazines, scissors, and glue. Have students cut out pictures of squares and triangles they find in the magazines and create collages.

Lacing Cards: Have students lace square and triangle shapes on lacing cards.

Geoboard Triangles: Have each student create 5 triangles on their geoboards and explain why each is a triangle.

Exercise 4 • page 79

Extend Play

Triangle Sort: Have students sort paper cutouts of equilateral, isosceles, and scalene triangles by lengths of sides.

Objectives

- Count 5 objects with one-to-one correspondence.
- Identify circles.

Lesson Materials

- Paper cups of different sizes
- Hula hoops and agility cones
- Stuffed toys or bean bags
- Circle Match Cards (BLM)
- Number Cards (BLM) 1 to 5
- Large, dull-tip needle threaded with yarn
- Squares of large-weave burlap
- Embroidery hoops
- Optional snack: banana or kiwi slices and triangle crackers

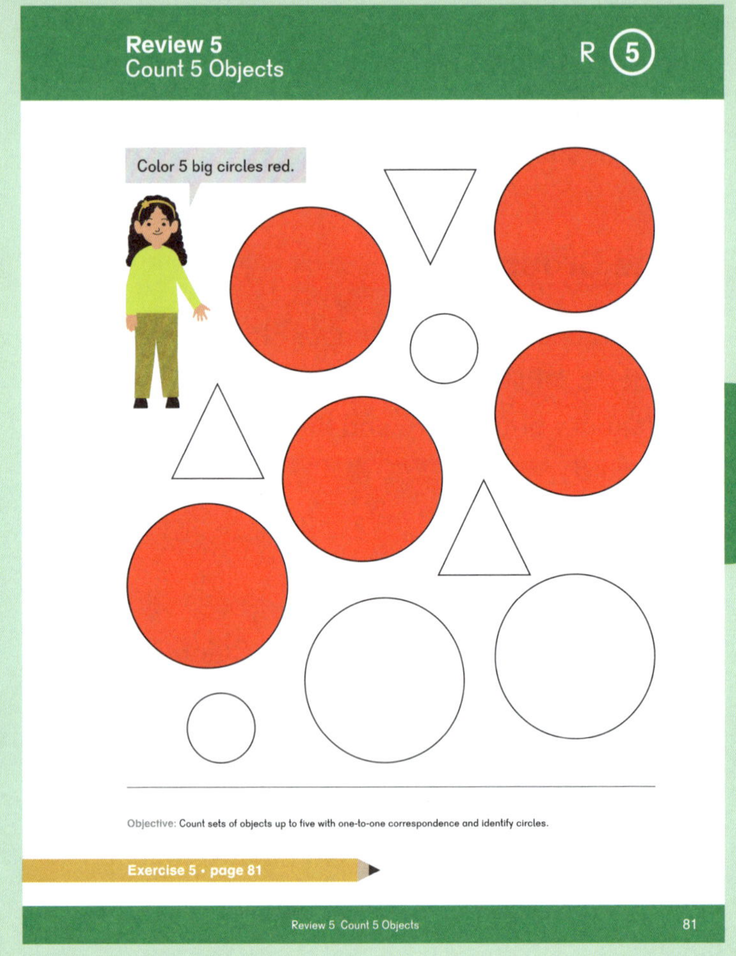

81

Explore

Have students find up to 5 examples of circles in the classroom and bring them to the group. Have them tell how many they found and hold up the matching Number Card (BLM).

Learn

Sing "Circle, Square, Rectangle, Triangle" (VR) with students. Draw each shape in the air with your finger as you sing about it and have students do the same. Have students look at page 81. Read Sofia's speech bubble and have students complete the task. Have students name the shapes not colored and give the reason for not coloring them.

Whole Group Play

Hula Hoop Toss: Put students in groups of 3. Two students hold a hula hoop in the air and the third tosses a bean bag or a stuffed toy through the hula hoop. Students take turns.

Hula Hoop Obstacle Course: Set up an obstacle course using hula hoops and agility cones.

Small Group Center Play

Circle Art: Have students dip the top of paper cups into paint and transfer the paint to art paper to create circle masterpieces.

Circle Match: Have students match Circle Match Cards (BLM) to Number Cards (BLM).

Exercise 5 • page 81

Extend Play

Burlap Sewing: Show students how to secure a square piece of burlap into an embroidery hoop, after they have named the shape of each. Provide each student with a needle threaded with yarn and a piece of burlap in a hoop. Teach them how to sew.

Objective

- Understand that an empty set is a set of zero.

Lesson Materials

- 0 Art Paper (BLM)
- Daubers of different colors
- Circle Match Cards (BLM)
- Number Cards (BLM) 0 to 5
- Geoboards and rubber bands
- Optional snack: nut-free trail mix

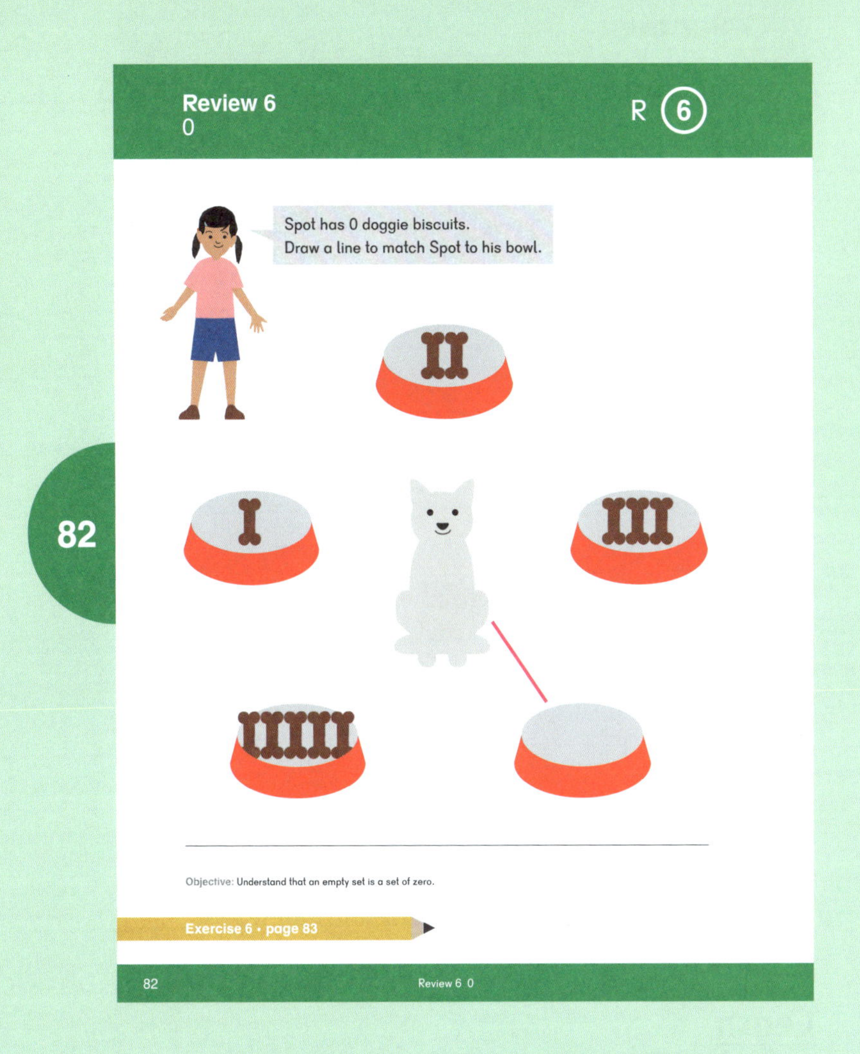

Explore

Have students work with a partner to name things of which there are 0 in the classroom.

Learn

Have students look at page 82. Tell students that the characters all named their dogs Spot. Have them tell what they notice is different about this Spot from the other Spot they met in **Dimensions Math® PKA**. Then have them tell how many spots Spot has.

Have students point to the dog bowl with 5 biscuits in it. Repeat for 3, 2, and 1. Read Mei's speech bubble to them and have them complete the task.

Whole Group Play

Play "Magic Thumb" counting up and back from 0 to 5.

Small Group Center Play

Zero Dots: Use 0 Art Paper (BLM) or create a large outline of a "0" on art paper, and invite students to use different colored daubers to decorate the 0. After they are done, ask them which color they did not use and tell you, "There are _____ (color) dots."

Matching: Have students match Circle Match Cards (BLM) to Number Cards (BLM), including blank cards and the 0 card.

Geoboards: Have students create shapes on their geoboards. When they are done, have them tell which shape of which there are 0.

Exercise 6 • page 83

Extend Explore

Story of 0: Have students tell stories about things of which there are 0 in the classroom or in their homes. They can either tell their story to a friend or record it. These stories could be illustrated and made into a class book with the dictated stories included.

Objectives

- Count with one-to-one correspondence.

Lesson Materials

- Number Cards (BLM) 1 to 10
- Number Cards — Large (BLM) 0 to 10
- Dominoes or Domino Cards (BLM), 1 to 10
- Play dough
- Lacing cards showing numerals 0 to 10, mixed up
- Plastic needles
- Yarn
- Optional snack: mini cheese balls and crackers

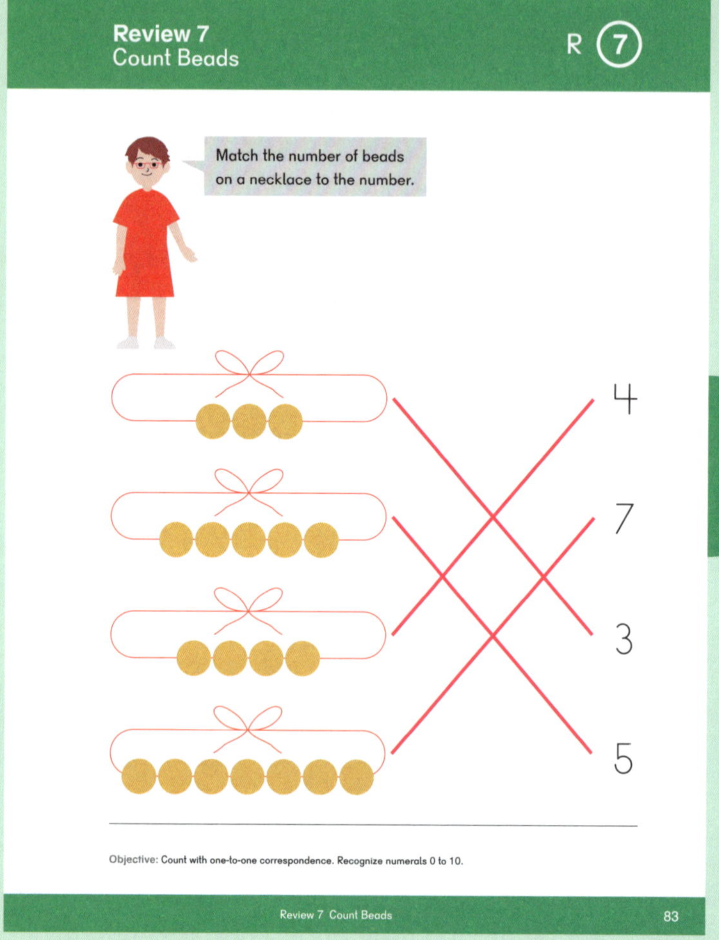

83

Explore

Assign each student a number, 0 to 10. Have them find the number in the classroom and point to it.

Learn

Give each pair of students a set of Number Cards (BLM) 0 to 10. Have them sort the cards into two groups with the reason for the sort being the appearance of the numbers. One group will be the "curved" group and the other group will be the "straight" group. 2, 5, 9, and 10 could lead to interesting discussions when students discuss why they put the numbers in one or the other group.

Have students look at page 83. Read Emma's speech bubble. Have students count the beads on the first necklace and draw a line to the 3. Have them complete the task.

Have students look at page 84. Read Emma's speech bubble. Have students count the beads on the first necklace and circle 5 of them. Have them complete the task.

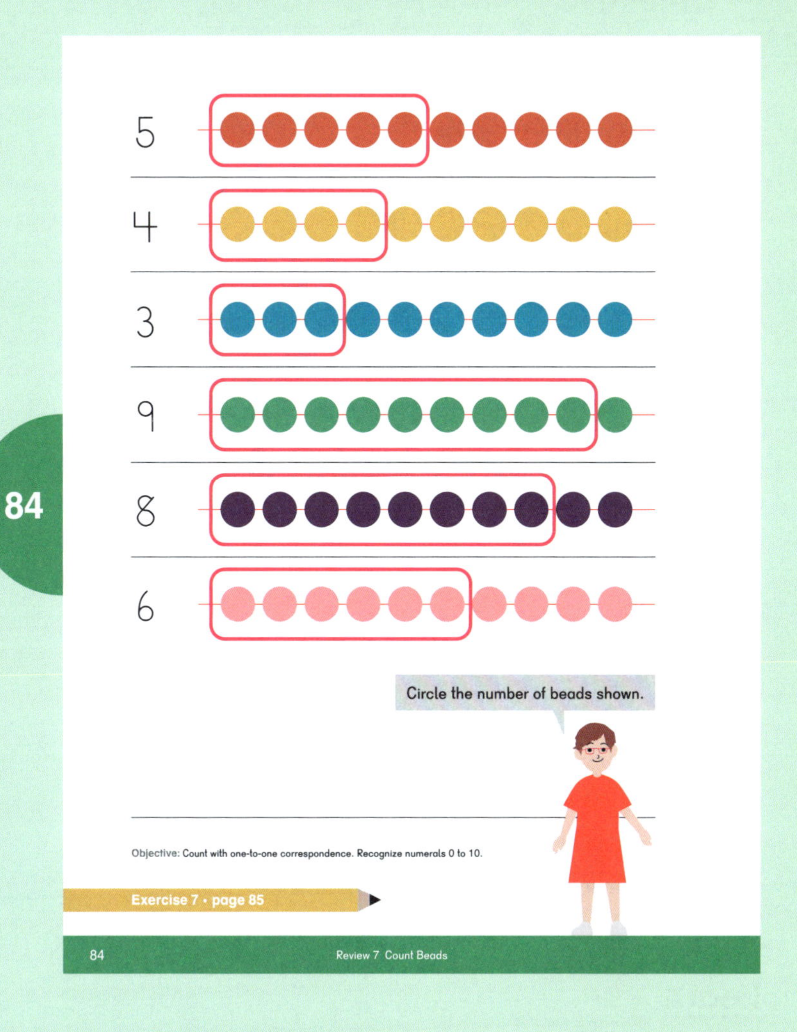

Whole Group Play

Flash and Fingers: Flash Number Cards (BLM) and have students show the correct number of fingers. Remind them to show a fist for 0.

Simon Says: Play Simon Says, having students do jumping jacks, windmills, etc., a certain number of times, including 0. Use Number Cards — Large (BLM) to show them how many times to do each exercise.

Small Group Center Play

Matching: Have students match dominoes or Domino Cards (BLM) to Number Cards (BLM).

Play Dough Numbers: Have students use play dough to create zeros, ones, and tens.

Number Lace-up: Using lacing cards, yarn, and dull needles, have students lace the numbers in order starting with 1.

Exercise 7 • page 85

Extend Play

Baking 1 and 0: Have students use dough to make zeros and ones. Bake and enjoy!

Objective

- Identify and extend patterns.

Lesson Materials

- Stickers of several types
- Linking cubes of several different colors
- Animal counters
- Building materials
- Tubs of sand (or access to sandbox)
- Optional snack: red and green grapes or 2 types of crackers to make patterns before eating

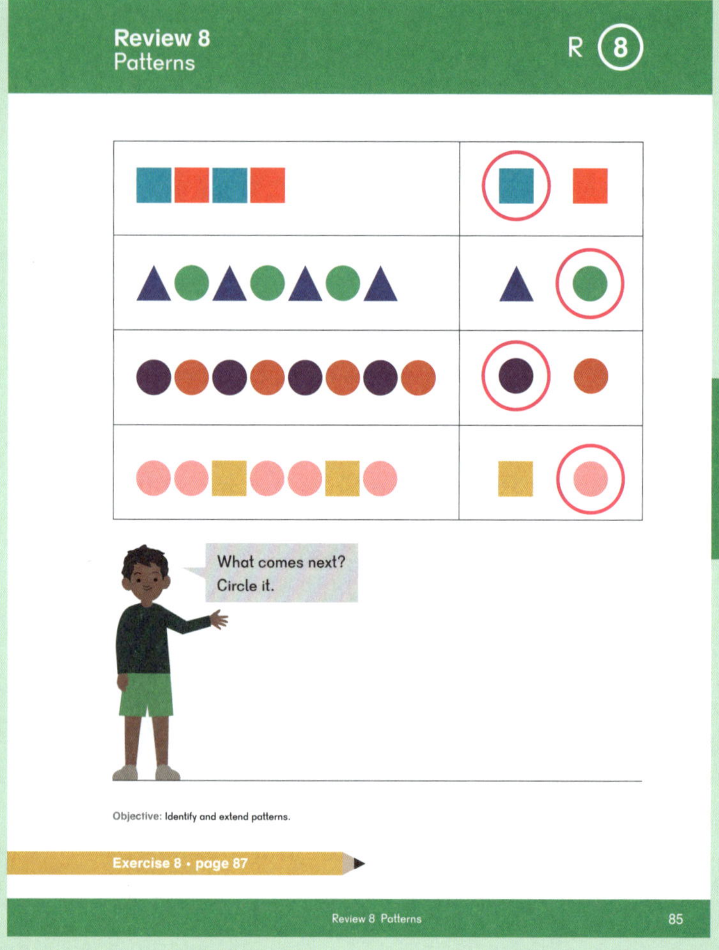

Explore

Have students find a repeating pattern in the classroom and point to it, naming the part that repeats. If necessary, suggest that students look at clothing and classroom decorations.

Learn

Have students use linking cubes to create trains of repeating patterns. Then have them exchange their trains with a partner and extend each other's patterns. Have students look at page 85. Read Alex's speech bubble and have them complete the task.

Whole Group Play

Lead children in movement and clapping patterns.

Small Group Center Play

Pattern Art: Provide stickers and crayons and have students create repeating patterns.

Pattern Build: Have students build structures using a repeating pattern in their building materials.

Sand Patterns: Have students use their fingers to draw repeating patterns in sand.

Tell Me a Story: Have students create repeating patterns using 8, 9, or 10 animal counters. Then have other students extend the patterns. Have students tell stories about the animals they used.

Exercise 8 • page 87

Extend Learn

Can You Make What I Made?: Have students work in pairs. One student describes a pattern that he or she made and the other student creates the same pattern.

Objective

- Compare objects by length.

Lesson Materials

- Craft sticks, 1 per student
- Paper strips of different lengths
- Linking cubes of several different colors
- Pieces of yarn or string of different lengths
- Xylophones
- Jump ropes of different lengths
- Optional snack: Various length cheese sticks

86

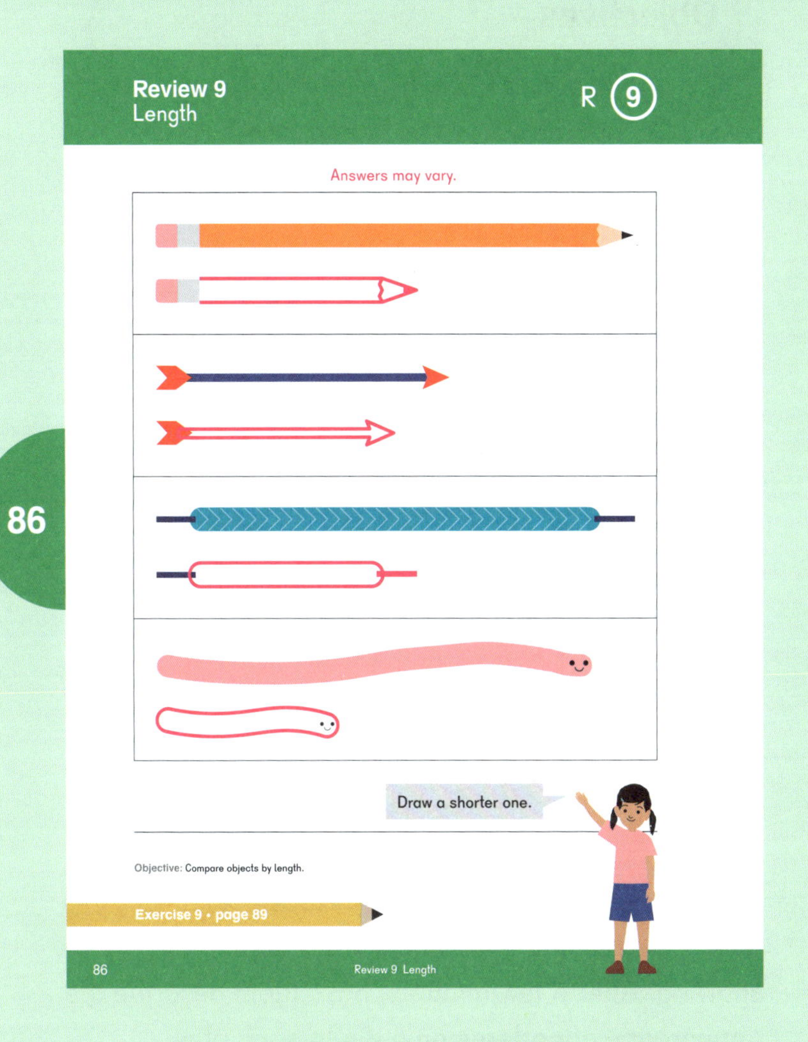

Explore

Give each student a craft stick. Have them look around the room to find objects which they think are longer than a craft stick. After looking and predicting, have them compare the actual length of their chosen object to a craft stick. Discuss student findings. Repeat for objects that are shorter than a craft stick.

Learn

Give each student a paper strip. Have them order the strips from shortest to longest, using the same start point for the bottoms of the strips. Then have students look at page 86. Read Mei's speech bubble and have them complete the task.

Whole Group Play

Create a train of linking cubes using a repeating pattern. Have students name the part that repeats. Then have them create a train that is longer than yours using a repeating pattern. After ensuring that a student's train is longer than yours, have him or her create a train that is shorter than yours.

Small Group Center Play

Sort: Provide different lengths of yarn or string for students to sort by length.

Xylophone Sounds: Have students play notes on a xylophone and compare the sounds made by the various lengths of the xylophone bars. Have students play patterns on the xylophones.

Jump Rope: Talk about the lengths of different ropes first, then head outside to jump.

Exercise 9 • page 89

Extend Explore

Measure and Draw: Have students measure the length of objects in the classroom using craft stick units. After measuring, have them draw a picture of the measured objects and draw lines to represent the number of craft sticks used to measure it.

Objectives

- Subitize dice patterns.
- Recognize quantities on ten-frame cards.

Lesson Materials

- Dice, 1 per pair of students
- Ten-frame Cards (BLM) 0 to 10
- Number Cards (BLM) 0 to 10
- Linking cubes
- Dot Cards (BLM) 0 to 10
- Square pieces of white paper
- Black crayons or markers
- Optional snack: square graham crackers covered with cream cheese and raisins in dice patterns

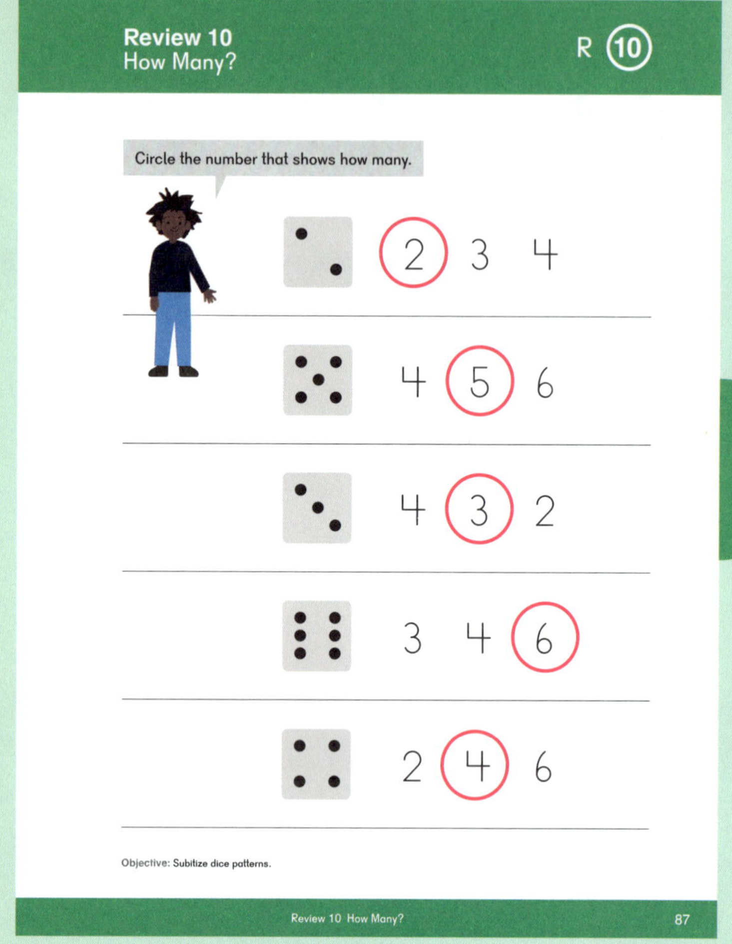

87

Explore

Give each pair of children a die. Have them take turns rolling the die and telling how many dots (pips) are showing. After a few minutes, have them describe the appearance of patterns on a die to each other.

Learn

Ask students to describe the patterns on a die. If necessary, give them prompts. For example, "For a 2, the dots are in corners of the square."

Have students look at page 87. Read Dion's speech bubble and have them complete the task.

Have students look at page 88. Read Sofia's speech bubble to them and have them complete the task.

Whole Group Play

Flash Ten-frame Cards (BLM) and have students call out, or show on their fingers, the numbers shown.

Small Group Center Play

Roll and Build: Have students roll a die and build a tower of linking cubes. For each roll, the student adds that many linking cubes to the tower. How tall can the tower get before it topples?

Matching: Have students roll a die, then find the number shown on a Number Card (BLM), a Dot Card (BLM), and a Ten-frame Card (BLM).

Draw the Pips: Have students use black crayons or markers and square pieces of paper to draw the faces of a die.

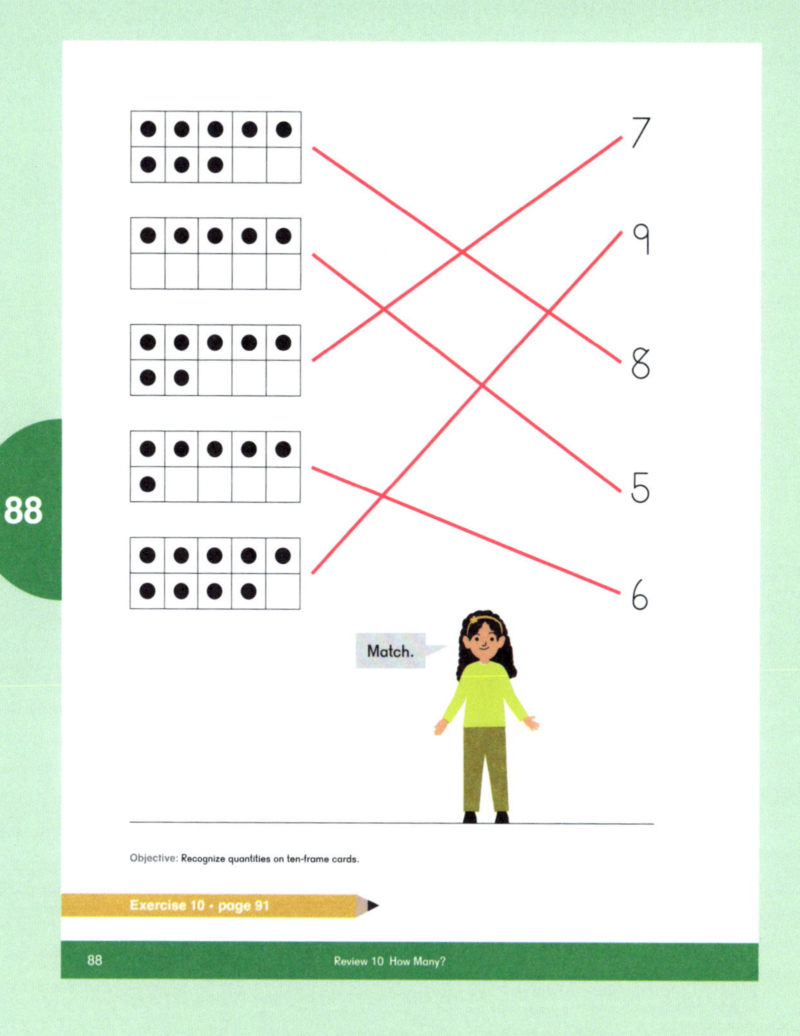

Exercise 10 • page 91

Extend Learn

Matching: Have students pair Ten-frame Cards (BLM) that together make a sum of 5 to 10.

Objective

- Identify first through fifth from a starting point.

Lesson Materials

- Small bags containing 5 counters, each of a different type, 1 bag per student
- Square tile or square paper cutouts of different colors
- Objects to be used as markers
- Musical instruments
- Optional snack: triangle crackers

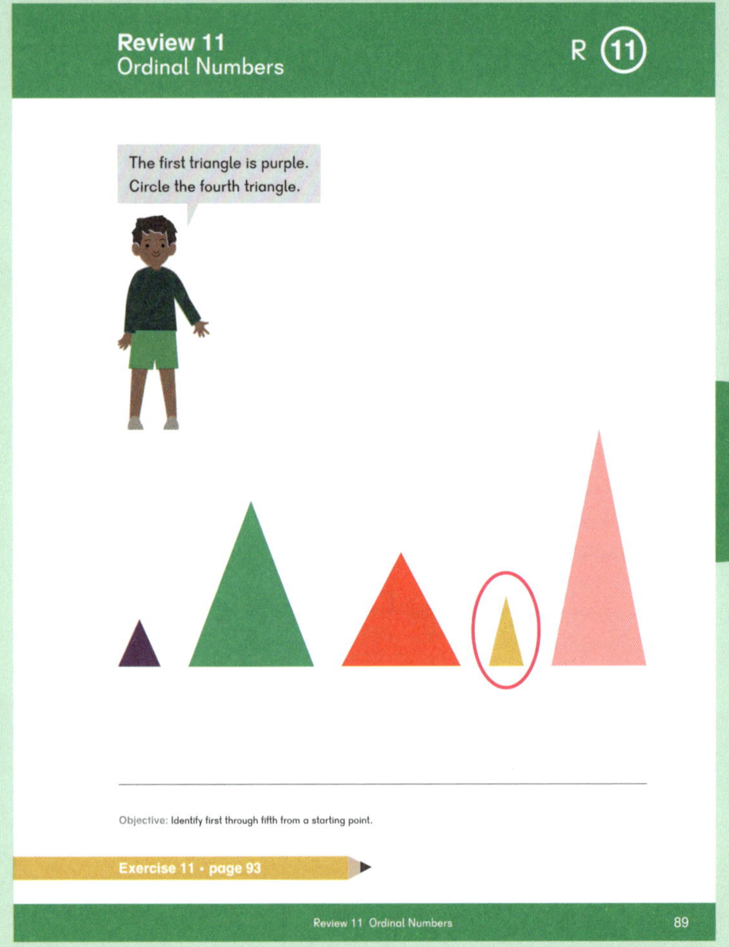

Explore

Give each student a bag of counters and have them line them up horizontally. Then have them work with a partner to say which counters are first through fifth from the left, then from the right.

Learn

Have students look at page 89. Read Alex's speech bubble. Have them put their finger on the purple triangle, then complete the task.

Whole Group Play

Have students race (toe-to-heel, duck walk, crab walk, etc.) and tell which of them came in first through fifth.

Small Group Center Play

Ordered Triangles: Have children draw a column of five triangles. Tell them to color the second triangle from the top a certain color. Allow them to decorate the other triangles as they wish.

Ordered Patterns: Have students use colored tiles or square paper cutouts of different colors to create patterns. Have them put a marker on the third square from the left.

Play 5 Notes: Have students use musical instruments to play 5 notes from low to high. Have them play the fifth note 5 times.

Exercise 11 • page 93

Extend Play

Mary Had a Little Lamb: Teach children "Mary Had a Little Lamb" as on page 116 of this Teacher's Guide.

Objectives

- Recognize spheres, cubes, and cylinders.
- Make a graph from a picture.
- Recognize circles, squares, rectangles, and triangles.

Lesson Materials

- Cubes, cylinders, and spheres from geometry sets and manipulative kits
- Real-life examples of the solids listed above, such as dice, soup cans, and different sizes of balls
- Square piece of paper
- Cardboard tubes, 1 per student
- Small pieces of fabric
- Rubber bands
- Aluminum foil
- Dry rice, dried beans, and/or sand
- Paper triangles, straws, pieces of pool noodle, and tubs of water to build boats
- Drums
- Optional snack: food in the shapes of cubes, cylinders, and spheres, such as cheese cubes, gelatin cubes, marshmallows, grapes, and puff cereal

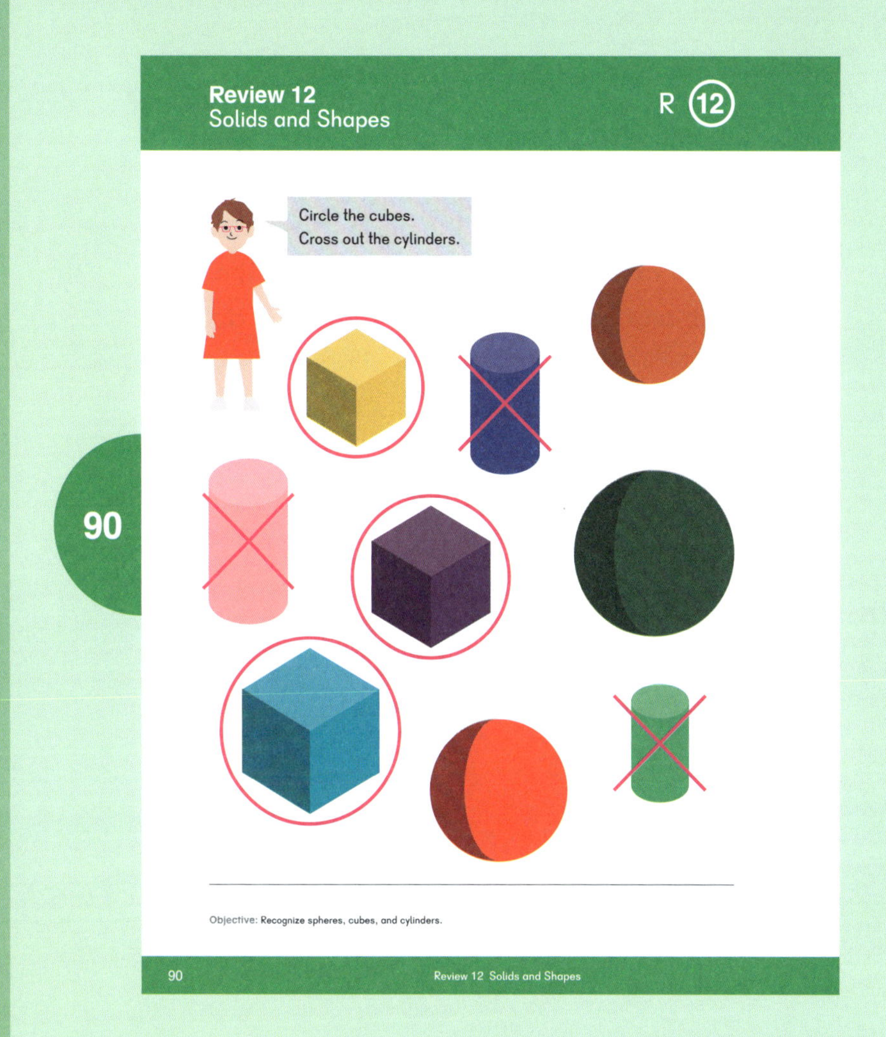

Explore

Give each small group of students the solids listed in **Materials** and have them sort them, explaining the reason for their sort. Have them sort them as many times as possible, each time explaining the reason for their sort.

Learn

Hold up a cube. Have students name it and tell how they know it's a cube. For example, "It has no curved surfaces. The face of a cube is a square. It has 8 points." Repeat for the sphere and the cylinder.

Hold up a square piece of paper and a cube. Ask students what is alike and what is different about them. Encourage them to use the words "flat" and "solid."

Have students look at page 90. Read Emma's speech bubble and have them complete the task.

Have students complete the task on page 91.

Have students look at page 92. Read Dion's speech bubble. Have students count the cubes and color in four of the rectangles above the cube. Have students complete the task. Ask which solid is shown the most on the page and how they can tell from the graph.

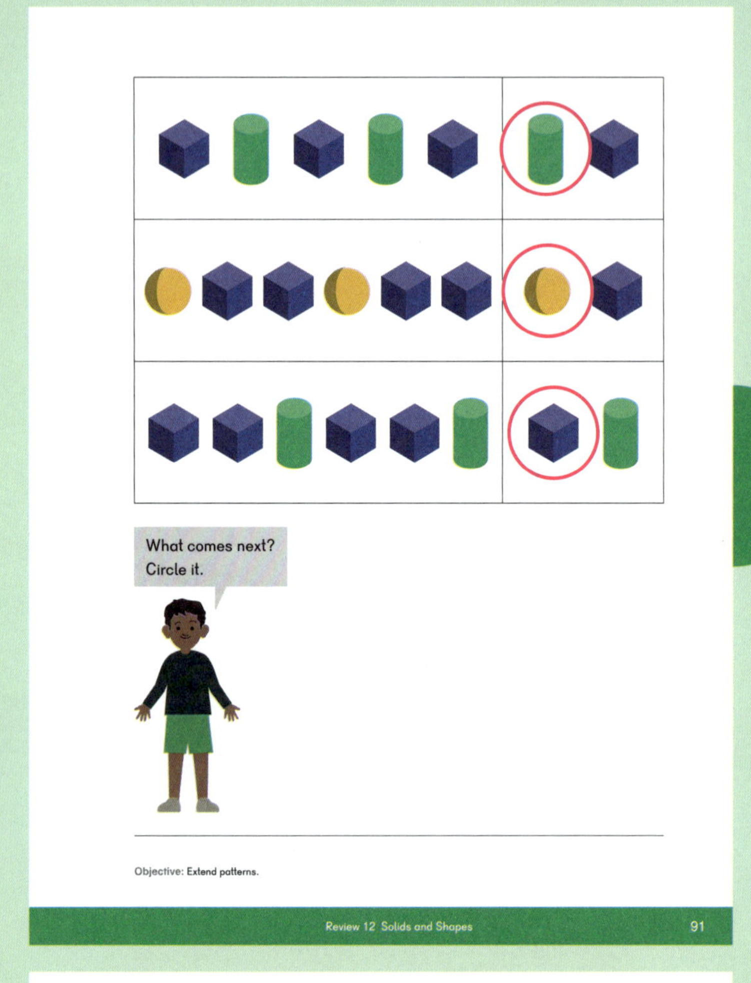

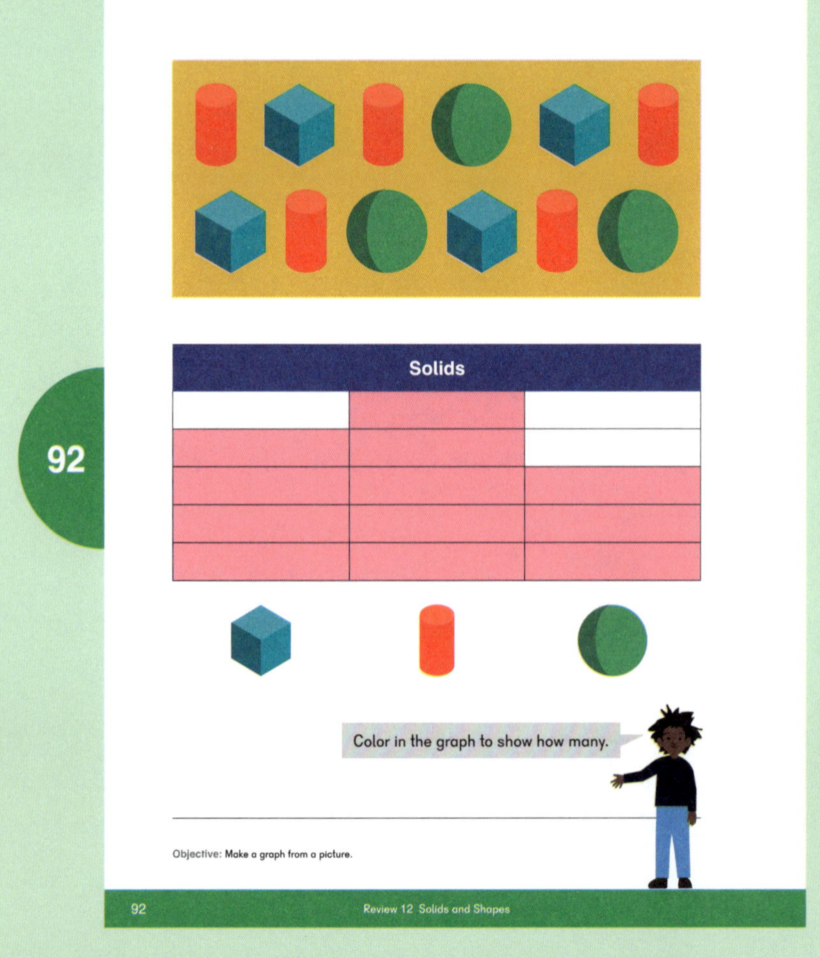

Have students look at page 93. Tell them that Mei is at Shapes Harbor. Ask them if they know what a harbor is, or if they can guess by looking at the picture. Read Mei's speech bubble. Display a yellow circle, green square, orange rectangle, and purple triangle to remind students of the directions. Have students complete the task.

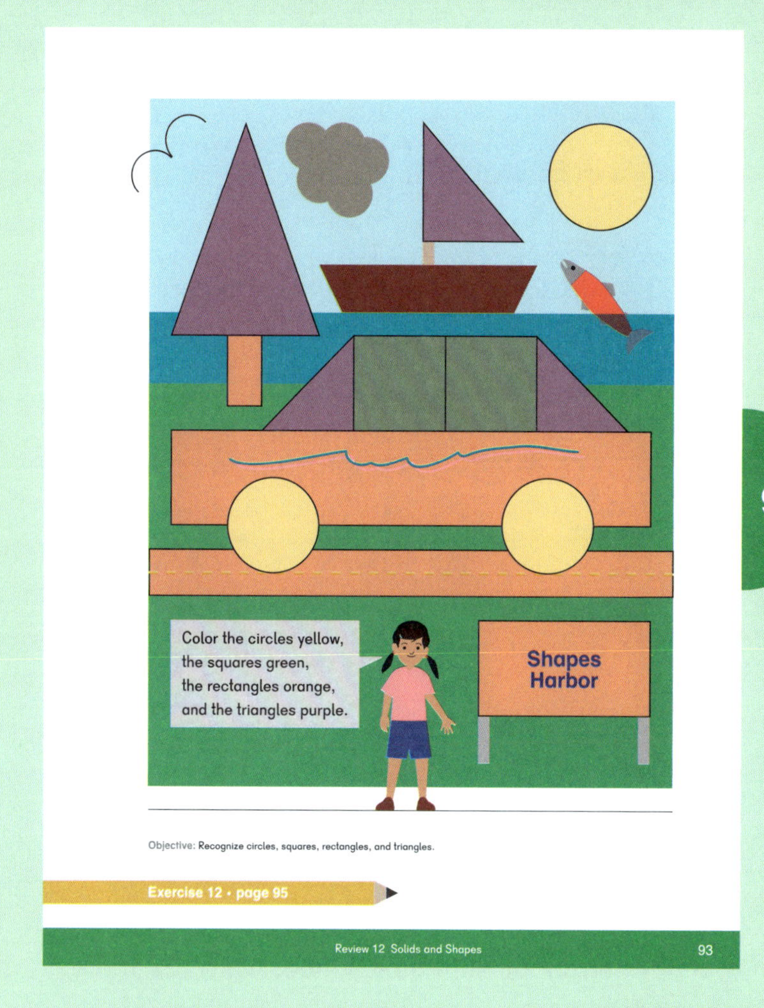

Whole Group Play

Roll It Over: Have students play in groups of four. Have them sit on the floor and roll a ball to each other. Have them repeat with a cylinder.

Sing "Row, Row, Row Your Boat."

Read *Shapes, Shapes, Shapes* by Tana Hoban.

Small Group Center Play

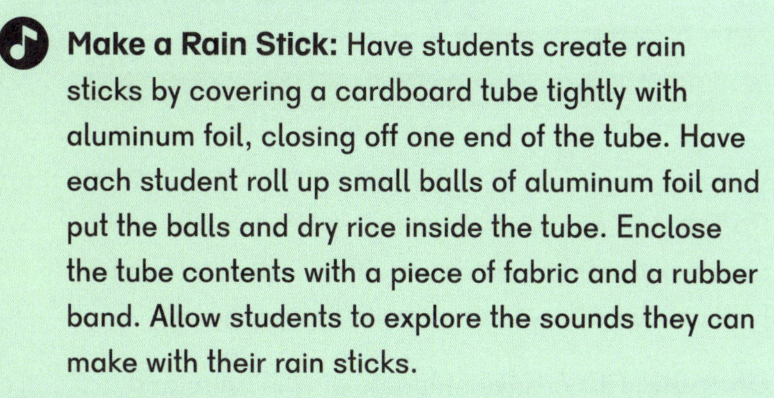

Make a Rain Stick: Have students create rain sticks by covering a cardboard tube tightly with aluminum foil, closing off one end of the tube. Have each student roll up small balls of aluminum foil and put the balls and dry rice inside the tube. Enclose the tube contents with a piece of fabric and a rubber band. Allow students to explore the sounds they can make with their rain sticks.

Build a Boat: Have students punch two holes in a paper triangle and put a straw through it. Then have them put the straw into a pool noodle piece. Have them put their boats in a tub of water and blow on the paper sails for boat races. Have them tell which boat came in first through fifth.

Cylinder Beats: Show students different drums that are shaped like cylinders. Have students beat a sound pattern on the drums.

Exercise 12 • page 95

Extend Play

Rain Sticks: Have students make more rain sticks, using longer or shorter cardboard tubes, or replacing the aluminum foil balls and rice with other items such as dried beans or sand. Discuss how these changes alter the sounds they can make with their Rain Sticks.

Review 13 Which Set Has More?

Objective

- Compare two sets of objects to find which has more.

Lesson Materials

- Two-color counters
- Plastic eggs or other small containers, 1 per student
- Chairs set up for Musical Chairs
- Dice, 1 per student
- Blank Five-frames (BLM), 1 per student
- Optional snack: celery sticks and carrot sticks, more than one of the other for each student

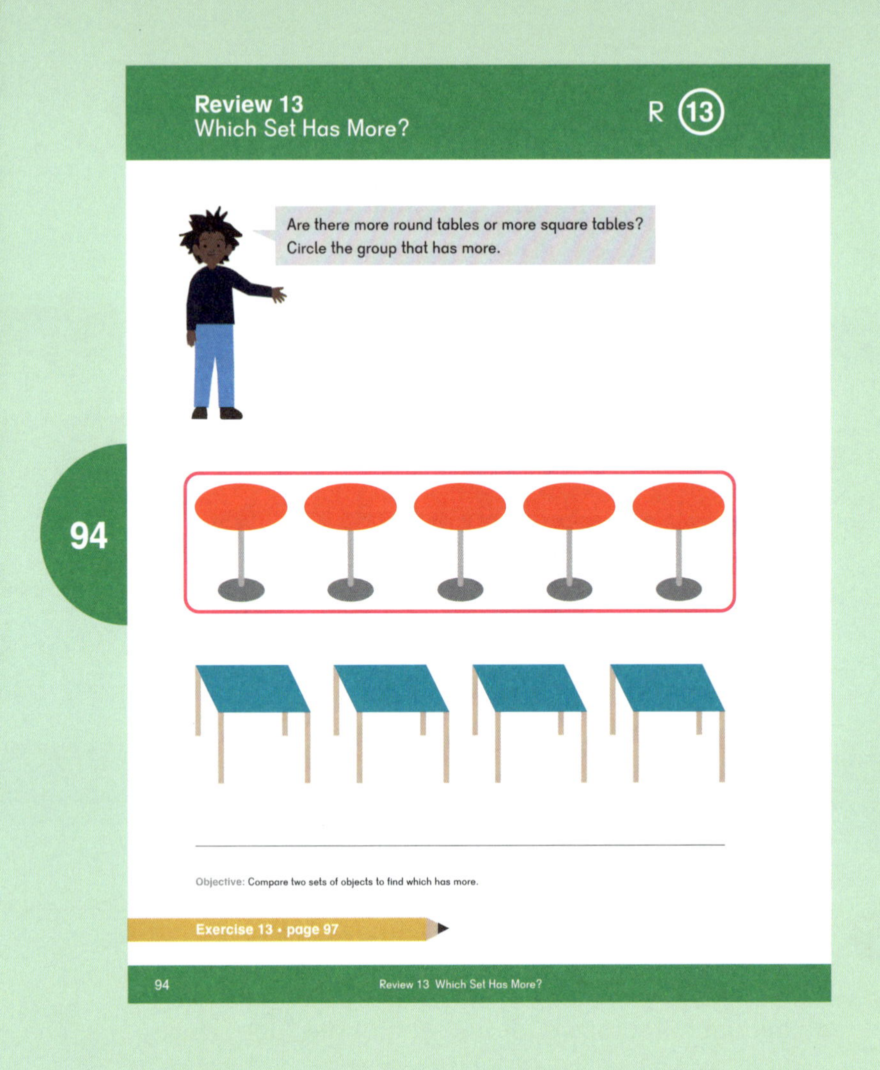

Explore

Give each student a small container and 5 two-color counters. Have them play "Shake and Pour" by pouring out the counters several times and talking to a partner about what they notice. Ask, "Are there more red or white counters?" Notice their problem-solving strategies. Have them shake and pour again, and tell a partner if they have more red or white.

Learn

Have students look at page 94. Read Dion's speech bubble and have them complete the task.

Whole Group Play

Musical Chairs

Small Group Center Play

Roll and Compare: Give each pair of students a Blank Five-frame (BLM), 2 dice, and 5 counters. Each student rolls a die and compares his number of pips with his partner's. The student with more pips showing puts a counter on his five-frame card. The first student to fill their card wins.

Patterns: Have students create patterns with counters or other classroom objects, then tell which type of object they used more in their patterns.

Dramatic Play: Have students play in pairs and dress up, taking turns being the one to put on more clothes.

Exercise 13 • page 97

Extend Play

Shake and Pour More: Play "Shake and Pour" with 7 or 9 two-color counters.

Objective

- Compare two sets of objects to find which has fewer.

Lesson Materials

- Bags containing 2 types of counters, 5 of each, 1 bag per student
- Chairs set up for Musical Chairs
- Dice
- Dominoes
- Drinking straws of two colors, cut into half-inch pieces
- Pipe cleaners
- Dot Cards (BLM)
- Optional snack: 2 pretzel sticks and 1 pretzel rod per student

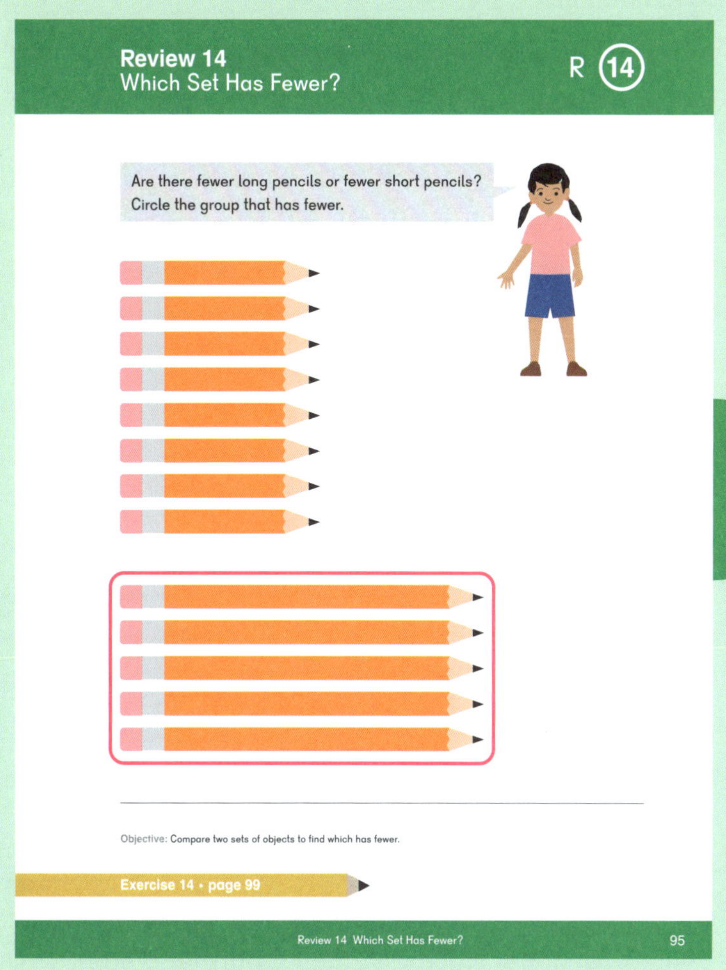

95

Explore

Give each student a bag of counters. Have them make two sets of counters so that there are fewer of one type of counter than the other, then compare. Ask, "How many different ways can you make two sets so that there are fewer ______ than ______?"

Learn

Have students look at page 95. Read Mei's speech bubble. Ask students how they think they can compare the two sets of pencils.

Whole Group Play

Musical Chairs

Small Group Center Play

Roll and Compare: This time, the student whose dice shows fewer pips gets a point.

Domino Dots: Lay 9 dominoes facedown for each pair of students. Each student turns over a domino. The student whose domino shows fewer dots keeps both dominoes. If their dominoes show the same number of dots, each student keeps 1 domino. The student with the most dominoes at the end of the game wins.

Pipe Cleaner Straw Art: Have students put drinking straw pieces on a pipe cleaner and tell which color they used fewer pieces of.

Extend Play

Dot Card Flash: Flash 2 Dot Cards (BLM) and have students point to the card showing fewer dots. For added challenge, have students tell how many fewer dots are on the card they point to.

Objective

- Identify the whole when two parts are given.

Lesson Materials

- Parts and Whole Mats (BLM), 1 per student
- Animal counters, 5 per student
- Paper plates, 1 per student
- Strips of brown paper
- Plastic eggs of different colors
- Domino Cards (BLM) or dominoes
- Number Cards (BLM)
- Linking cubes or Lego® blocks
- Optional snack: orange slices and apple slices for students to use in telling put together stories before eating

Explore

Give each student a Parts and Whole Mat (BLM) and animal counters. Have them tell a partner a story about 5 animals in two parts, then model the story parts by placing the counters in the parts on the mat. Have students say, "___ and ___ make ___," as they move the counters from the two parts to the whole on the mat. Ask them how many different ways the 5 animals can be in two parts and have them take turns telling stories and modeling them.

Learn

Have students look at page 96. Read Dion's speech bubble. Have students tell the two parts on the first pair of leaves, then say, "2 and 1 make 3," and draw a line to the leaf showing 3 ladybugs. Repeat for the other pairs of leaves on the left side of the page.

Have students look at page 97. Ask them how many frogs are not on the lily pad and how many frogs are on the lily pad in the first row. Then have them say, "3 and 1 make 4," and circle the 4. Repeat for the other problems.

Whole Group Play

Make 5 and Sit!: Play music. While the music plays, students dance. When the music stops, students get into groups of 5 as quickly as possible and sit together. After sitting, students in each group think of a way to sort themselves into two parts (light hair, dark hair, etc.), then say the number sentence for their parts and whole.

Small Group Center Play

Bird Nests: Have students create bird nests by gluing brown strips of paper to a paper plate. Have them choose plastic eggs of two colors to put into their nests, up to 5 eggs in a nest. Then have them tell their partner a number sentence about the eggs in their nest, for example, "2 yellow eggs and 2 pink eggs make 4 eggs altogether."

Domino Match: Provide pairs of students sets of Domino Cards (BLM) or real dominoes. Have students find dominoes that match. A match is two dominoes that show the same whole. The parts can be the same or different. The student with the most matches at the end wins.

Building: Have students use 5 linking cubes or Lego® type blocks to create two structures. Have them match Number Cards (BLM) to the two parts. Then have them join the parts and match a Number Card (BLM) to the number of cubes or blocks used in all.

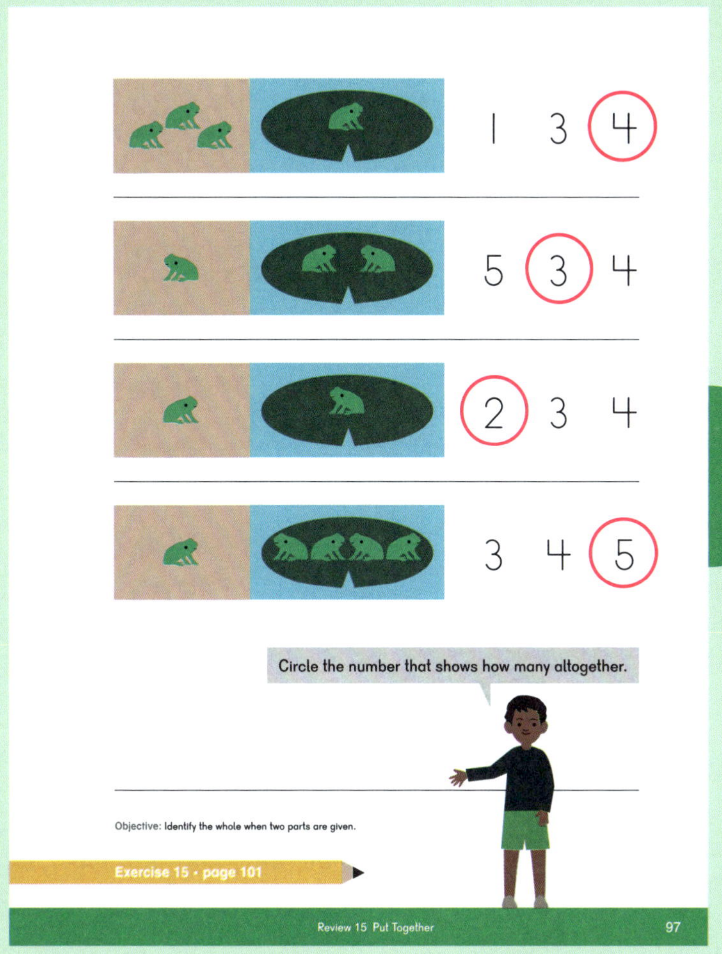

Exercise 15 • page 101

Extend Learn

Count by Twos: Have students line up 10 counters in columns of 5. Practice counting by twos with them. Then have them count the dots on the ladybugs on the page by twos.

DUPLO® and Lego® are trademarks of the Lego Group of companies.

Objective

- Subtract numbers within 5.

Lesson Materials

- Animal counters, 5 per student
- Paper cups, 1 per student
- *Stellaluna* by Janell Cannon
- Linking cube towers of 5, in pairs
- Counters
- Number Cards (BLM) 1 to 5
- Capes and toy bats
- Optional snack: fruits that bats like, such as mangoes, bananas, and avocados

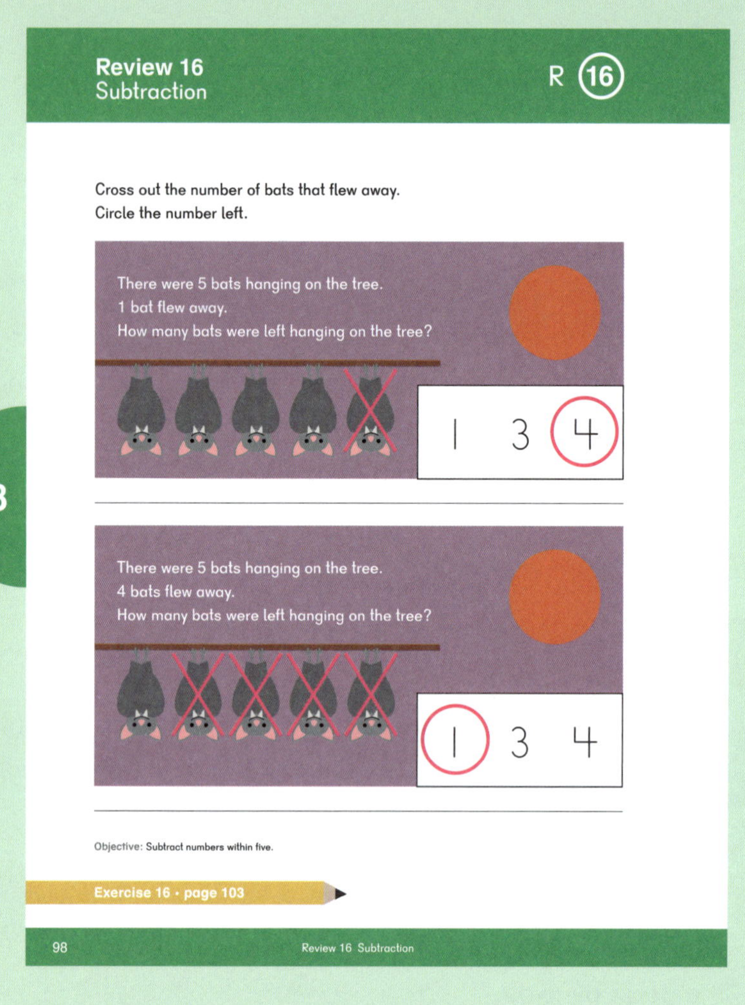
98

Review 16
Subtraction
R 16

Cross out the number of bats that flew away.
Circle the number left.

There were 5 bats hanging on the tree.
1 bat flew away.
How many bats were left hanging on the tree?
1 3 4

There were 5 bats hanging on the tree.
4 bats flew away.
How many bats were left hanging on the tree?
1 3 4

Objective: Subtract numbers within five.

Exercise 16 • page 103

98 Review 16 Subtraction

Explore

Give each student 5 animal counters. Tell a take away story about animals and have students model the story with their counters by removing the part taken away and hiding it under a paper cup.

Have students work in pairs. One student tells a take away story and the other models it with counters. Students take turns telling stories and modeling stories.

Learn

Teach students "Five Little Bats" (you can find this poem on page 116 of this Teacher's Guide.) After they have heard the rhyme several times, at the end of each verse, have students say, "There were ___ bats. One bat flew away. There are ____ bats left."

Have students look at page 98. Read the directions and have students complete the task.

Whole Group Play

Read *Stellaluna* by Janell Cannon, or another book about bats. Have students act out the story.

Small Group Center Play

Snap

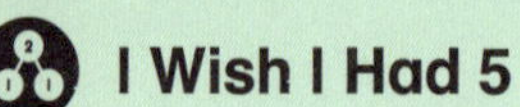
I Wish I Had 5

Tell Me a Story About Bats: Provide capes and bats and have students pretend to be Batwoman, Batman, or other superheroes of their choice.

Extend Learn

Helpful Bats: Tell students that bats eat many insects which would harm the crops that farmers raise. Some bats also pollinate flowers like bees. Have students draw a picture of bats being helpful to humans.

Objective

- Use logic to determine the sequence of events.

Lesson Materials

- Growing plants at different stages
- Strawberries, clean with stems attached, 1 per student
- Crayons, 1 red and 1 green per student
- *Goldilocks and the Three Bears*
- Optional snack: strawberries

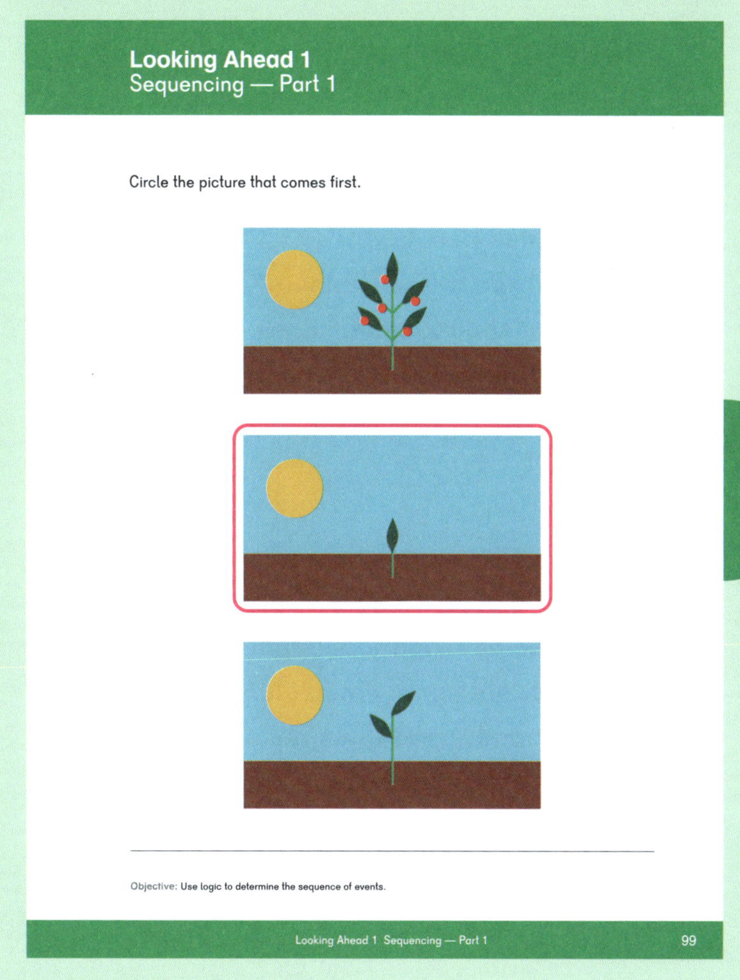

Explore

Have students look at the three plants and tell what they notice. Ask which plant was planted first, which was planted second, and which was planted third.

Learn

Give each student a strawberry, a red crayon, a green crayon, and a piece of art paper. Have them draw the strawberry on the left side of the paper. Have them each take one bite out of the strawberry and draw a picture of the bitten strawberry in the middle of the paper. Have them take another bite of the strawberry and draw a picture of what is left on the right side of their paper. Then ask them to tell which picture shows how the strawberry looked at first, and which shows how the strawberry looked last.

Have students look at page 99. Read the directions, then ask students which of the pictures shows the plant when it first started to grow. Have them complete the task.

Whole Group Play

Read *Goldilocks and The Three Bears*. Have students recall sequences from the story. For example, which chair did Goldilocks sit in first, second, and third?

Small Group Center Play

Free center day

Extend Play

Change the Ending: Tell students to change the ending of the story by predicting what would have happened if Goldilocks' parents came into the house of the 3 bears before the bears came home. They can either illustrate or record their stories.

Objective

- Use logic to determine the sequence of events.

Lesson Materials

- Magazines to cut up
- *Snow White* by the Brothers Grimm
- Optional snack: apple slices

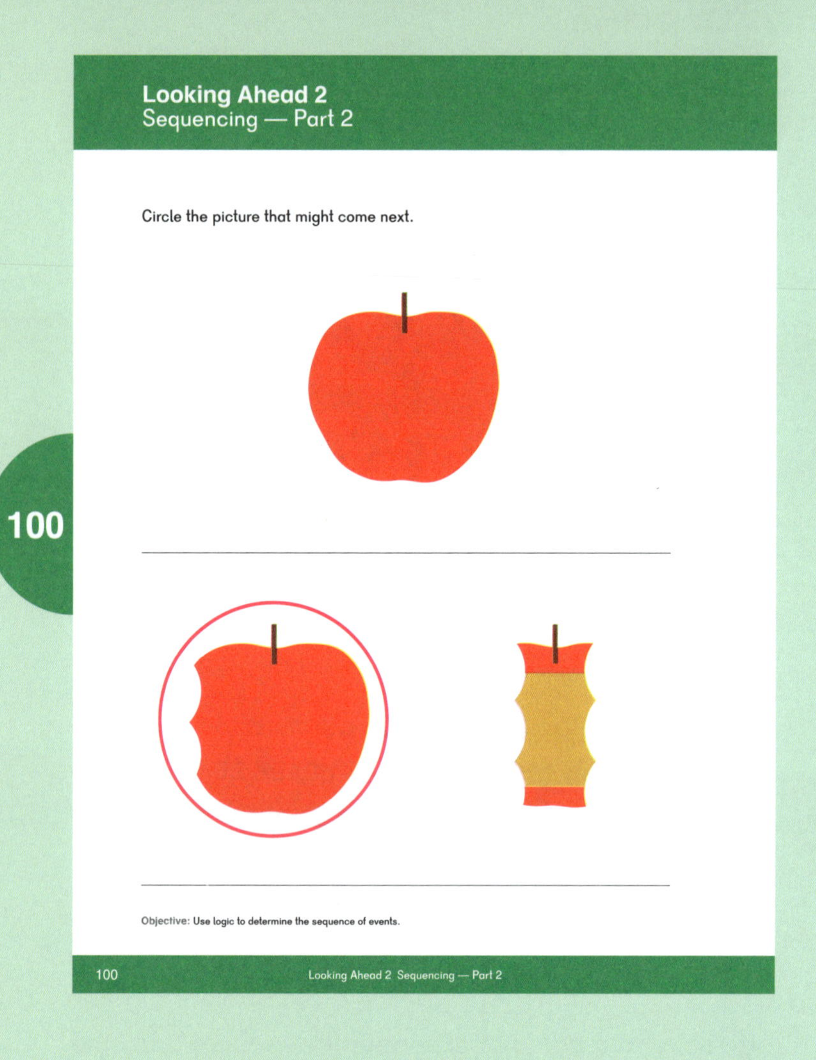

Explore

Give each student a magazine, scissors, a piece of art paper, and a glue stick. Tell them to find pictures in the magazine of things that happen in order. For example, a student might find pictures of a baby, a child, and an adult. Have them cut out the pictures and glue them to the paper from left to right in order. Have students share their findings with the group.

Learn

Have students look at page 100. Read the directions. Ask students which of the apples shown on the bottom of the page would come after the whole apple. Have them complete the task.

Have students look at page 101. Ask students what is happening on the page. Read the directions and have them complete the task.

Whole Group Play

Read *Snow White* by the Brothers Grimm. You may choose to read a family-friendly version of the story. Have students recall events from the story in order.

Small Group Center Play

Free center play

Extend Learn

Applesauce: Make applesauce with the students. Discuss the sequencing of the steps to create the recipe. Enjoy!

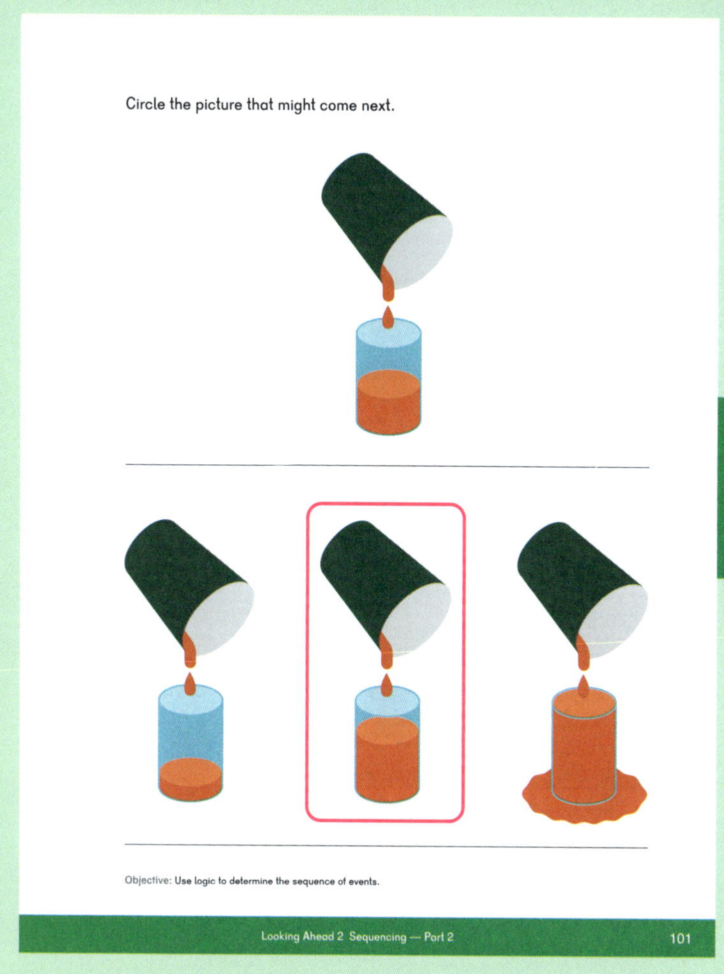

101

Objective

- Categorize objects by use.

Lesson Materials

- Numerous everyday objects that go together in groups of at least 2, such as, shoe and sock; pencil, sharpener, and eraser; T-ball, T-ball bat, soccer ball, small bouncy ball, and basketball
- Optional snack: grapes and raisins (Ask students, "Why do they go together?")

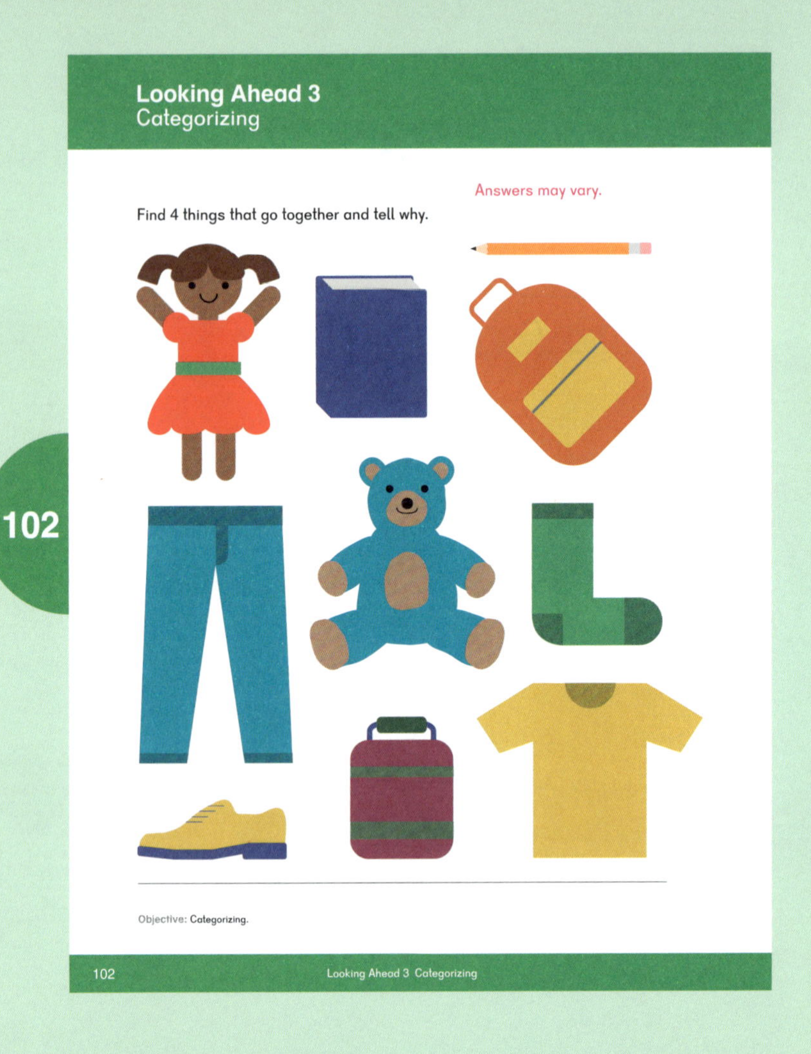

Explore

Show students an example of three objects that go together, such as a fork, a knife, and a spoon. Have them work in pairs to find three objects in the classroom that go together. Have them share their findings with the group, explaining why the objects go together.

Learn

Show students items such as those listed in **Materials** and identify each of them. Have students think of four objects from the materials that go together. The goal is to start a discussion as to why some objects go with other objects in certain cases. For example, a student may say that all of the balls go together because they are all balls. Another student may say that the T-ball and bat go with the soccer ball and basketball because they are all used for sports. Encourage students to categorize objects in many ways.

Have students look at page 102 and identify the objects that go on the page. Read the directions and have students discuss their ideas.

Whole Group Play

Play "I Spy" using things that go together as part of the clue. For example, "I spy something that goes with my chair." Have students volunteer answers, explaining their logic each time.

Small Group Center Play

Free center day

Extend Learn

5 Go Together: Have students find 5 objects that go together from the materials used in **Learn** and explain their logic.

Objective

- Add two parts to find the whole.

Lesson Materials

- Parts and Whole Mats (BLM), 1 per student
- Counters, 5 per student
- Number Cards — Large (BLM) 2 to 5, 1 set per student and extra sets for centers
- 4 fly swatters
- Addition and Equal Symbol Cards (BLM)
- Addition Facts Cards (BLM)
- Modified dice with sides labeled: +1, +1, +1, +2, +2, +2
- Number Cards (BLM)
- Optional snack: fish crackers

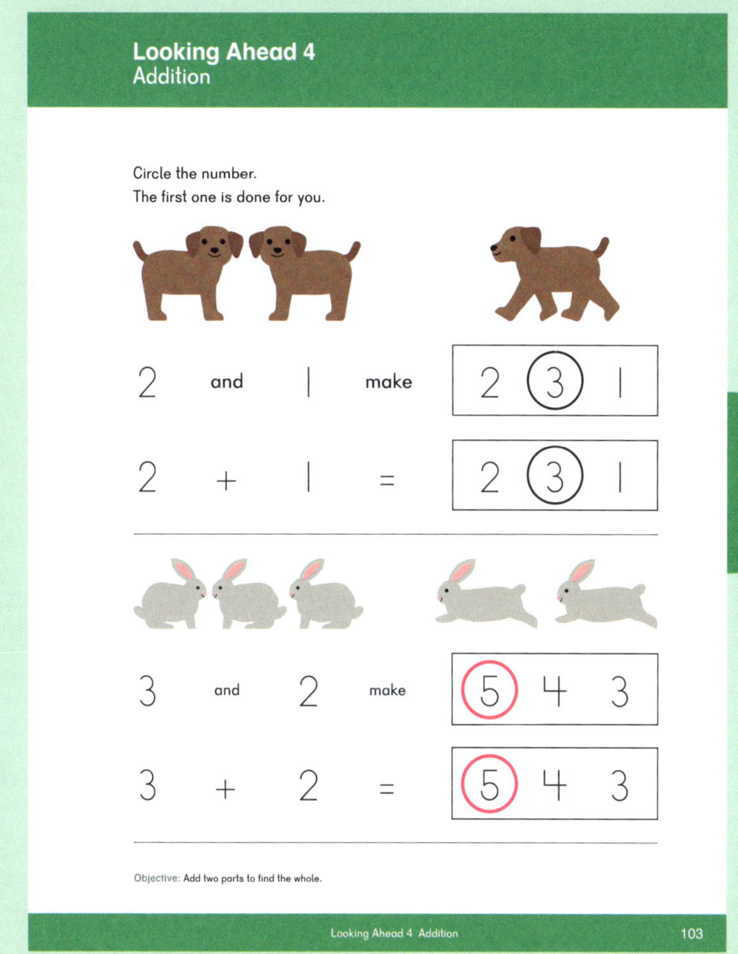

103

Explore

Give each student a Parts and Whole Mat (BLM) and 3 counters. Have them use their counters to make 3 in different ways and tell a partner a number sentence each time they do.

Give each student 1 more counter and have them repeat the activity for a whole of 4. Repeat for a whole of 5.

Learn

Have 3 students stand up. Tell a number story about them with parts 2 and 1. Have students say, "2 and 1 make 3," as you write, "2 and 1 make 3" on the board. Have the students sit down. Beneath what you wrote, write "2 + 1 = 3." Read each number sentence aloud. Point to the numbers, words, and symbols as you read them. Ask students what is alike and what is different about the two number sentences.

Point to the addition symbol and tell them that the symbol means put together and is read "plus." Point to the equal symbol, tell them that the symbol means "is the same as," and is read, "equals." Have them read both of the number sentences with you.

Erase the board. Have 4 students stand up. Tell a number story with parts 2 and 2. Repeat the procedure above. Erase the board.

Have 4 different students stand. Tell a number story with parts 3 and 1. Repeat the above procedure, but draw a box after the word "make" and "=." Have students tell you what number to write in the box.

Have students look at page 103. Read the directions and ask a student to tell a number story about the dogs. Have students read the two number sentences with you.

Ask a student to tell a number story about the rabbits. Have students read the first number sentence with you and tell you what number should be circled after "make." Repeat with the second number sentence.

Have students look at page 104 and repeat the procedure with the fish and tomatoes.

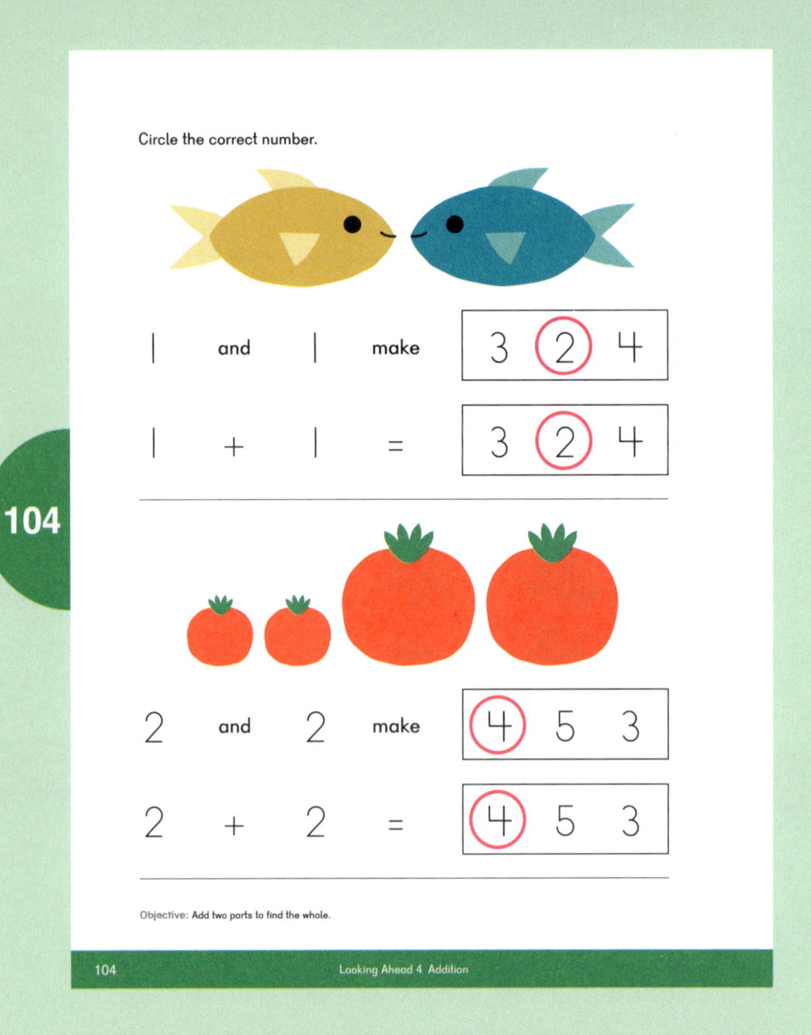

Whole Group Play

Matamoscas: Have students sit in a circle. In the middle of the circle, have four sets of Number Cards — Large (BLM) 2 to 5 faceup on the floor.

Call on four students. Give each student a fly swatter. Hold up an Addition Fact Card (BLM). Encourage all students to find the whole. If necessary, suggest they use their fingers to help them. The standing students must find a card showing the whole and swat it. They then pass the fly swatter to another student. Repeat with different addition fact cards.

Small Group Center Play

Make a Number Sentence: Set out Number Cards — Large (BLM) 2 to 4, facedown. Students choose a number card and roll a die, modified as directed in **Materials**. First, they model the two parts using counters. Then they create an addition sentence using their set of Number Cards (BLM) and the Addition and Equal Symbol Cards (BLM).

 **I Wish I Had 5**

Extend Learn

Tell Me a Story (With a Twist): Have students tell addition stories to a friend or record the stories, then use Number Cards (BLM) and Addition and Equal Symbol Cards (BLM) to make the accompanying number sentence.

Objective

- Take away one part from the whole to find the other part.

Lesson Materials

- Parts and Whole Mats (BLM), 1 per student
- Counters
- Paper cups, 1 per student
- Subtraction and Equal Symbol Cards (BLM)
- Number Cards (BLM)
- Linking cube towers of 5, in pairs
- Optional snack: "turtles" made of kiwi slices (body) and green grape halves (head and legs)

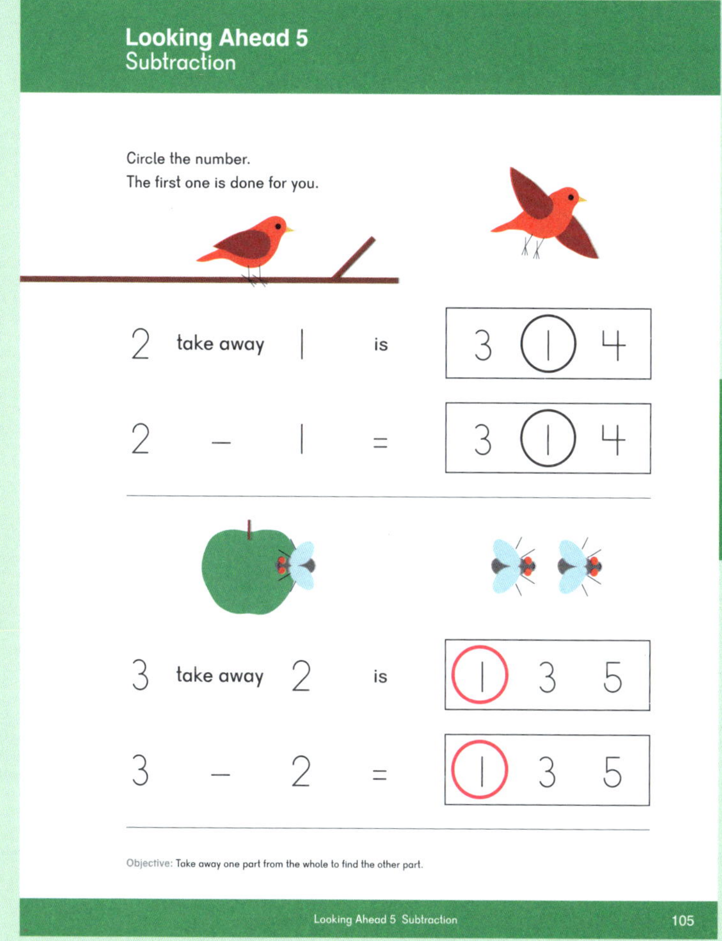

Explore

Give each student a paper cup and 3 counters. Have them play "Shake and Pour" (as described in Review 13 on page 139 of this Teacher's Guide) with a partner, hiding some of the counters under the paper cup. The partner must tell how many counters are hidden. Students take turns being the Pourer and the Teller.

Learn

Give each student a Parts and Whole Mat (BLM) and 3 counters. Tell a take away story for 3 – 1. Have students start by putting the counters in the whole of their mat. When you say that 1 was taken away, have them move 1 counter from the whole to a part. Write, "3 take away 1 is 2" on the board. Say the sentence, pointing to the numbers and the words as you do so, and have students repeat it. Beneath what you wrote, write "3 – 1 = 2." Point to the subtraction symbol. Tell students that in the stories they will be working with in Preschool, the symbol means "take away." Ask them what the equal symbol means.

105

Have students look at page 105. Read the directions. Ask a student to make up a story about the birds. Have students read the number sentences with you.

Ask a student to make up a story about the flies. Read the first number sentence and have students tell you which number should be circled. Have them circle the 1. Repeat with the second number sentence.

Repeat the procedure on page 106.

Whole Group Play

Hide and Seek

Small Group Center Play

 **Snap**

 **I Wish I Had 5**

Make a Number Sentence: Repeat the procedure from page 149 of this Teacher's Guide.

Extend Learn

Tell Me a Story (With a Twist): Have students tell take away stories to a friend or record the stories, then use Number Cards (BLM) and Subtraction and Equal Symbol Cards (BLM) to make the accompanying number sentence.

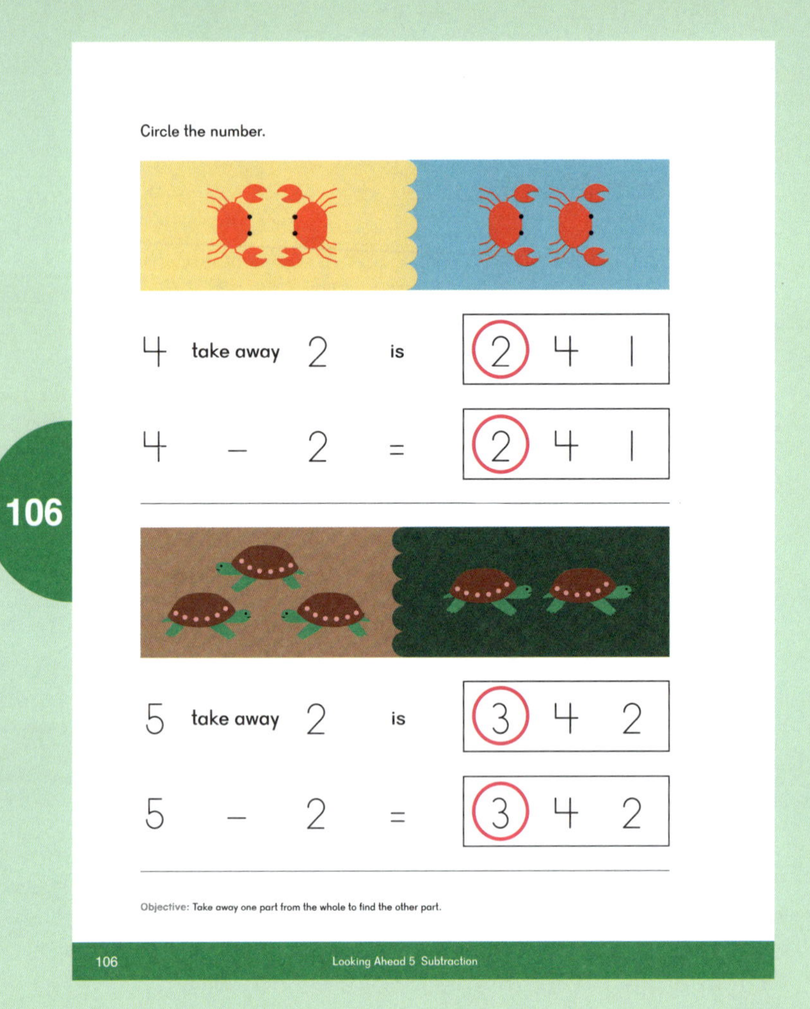

106

Circle the number.

4 take away 2 is (2) 4 1

4 − 2 = (2) 4 1

5 take away 2 is (3) 4 2

5 − 2 = (3) 4 2

Objective: Take away one part from the whole to find the other part.

106 Looking Ahead 5 Subtraction

Objective

- Draw vertical, horizontal, curved, and diagonal lines.

Lesson Materials

- Tubs of wet sand
- Play dough
- Optional snack: cheese sticks, square crackers, and round crackers

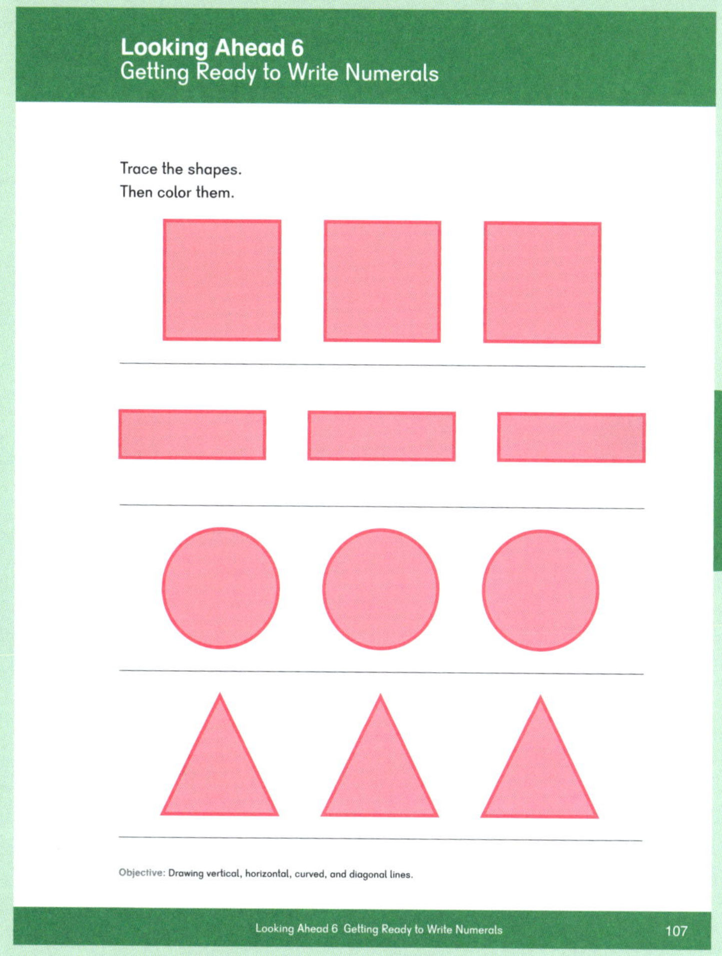

Looking Ahead 6
Getting Ready to Write Numerals

Trace the shapes.
Then color them.

Objective: Drawing vertical, horizontal, curved, and diagonal lines.

Looking Ahead 6 Getting Ready to Write Numerals 107

107

Explore

Ask students what the words "straight" and "curved" mean. Introduce the words "vertical," "horizontal," and "diagonal." Draw examples of each type of line on the board.

Have students find and share where they see the different kinds of lines in the classroom.

Learn

Have students look at page 107. Read the directions and have students complete the task.

Whole Group Play

Make a Line Simon Says: Show students your index finger. Draw a straight line in the air with your finger and have students do the same. Repeat with curved and diagonal lines. Play Simon Says using the types of lines as directed by Simon.

Small Group Center Play

Sand Drawing: Have students draw examples of the three types of lines in wet sand.

Play Dough Lines: Have students create examples of the three types of lines using play dough.

What Does It Look Like?: Have students describe the numbers 1 to 10 using the words "straight," "curved," and "diagonal."

Extend Explore

Lines and Numbers: Have students describe the numbers 1 to 10 using all of the vocabulary in this lesson.

Objective

- Find the number of letters in simple words.

Lesson Materials

- Cards with students' names on them
- Word Cards (BLM), 4 sets
- Number Cards — Large (BLM) 3 and 4
- 4 fly swatters
- Number Cards (BLM) 3 and 4
- Picture Cards (BLM) 3 and 4
- Word-Picture Cards (BLM)
- Optional snack: hard-boiled eggs

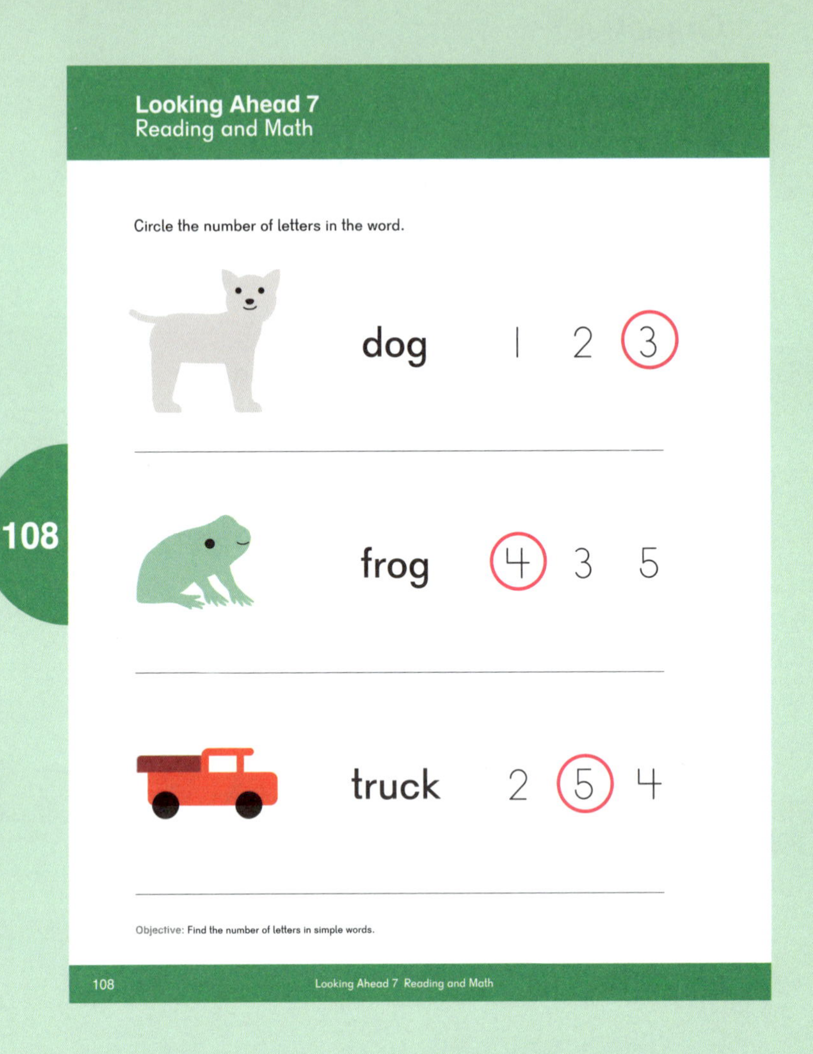

Explore

Put cards with students' names on them on the floor or a table. Have students find the card with their names and bring them to group.

Learn

Have students count the letters in their names. One at a time, call on a student. Have them show the card, tell how they knew the card showed their name, and tell the number of letters in their name. Have students help you make a graph showing the numbers of letters in students' names.

Have students look at page 108 and identify the objects on the page. Read the directions and ask students to say the letters in "dog." Then ask them how many letters there are. Have them circle the 3. Repeat for "frog" and "truck."

Whole Group Play

Matamoscas: Have students sit in a circle, in the middle of which are four sets of Word Cards (BLM). Call on four students, have them stand, and give each a fly swatter. Hold up a Number Card — Large (BLM) 3 or 4. The standing students must find a card showing a word with that many letters and swat it. Each of them will then hold up their word and the group will try to sound out the word. Provide prompts as needed. Repeat with four different students.

Small Group Center Play

Match It!: Set out Picture Cards (BLM), Word Cards (BLM), and Number Cards (BLM) 3 and 4. A match is one of each.

Draw It!: Have a set of Word-Picture Cards (BLM) facedown, crayons, and art paper at the center. Students pick a card from the top of the stack, then draw a picture of the word.

Extend Play

Draw It! (With a Twist): Set out Word Cards (BLM) facedown, crayons, and art paper. Students draw a card, sound out the word, and draw a picture of the word.

Exercise 1 • pages 73–74

Exercise 2 • pages 75–76

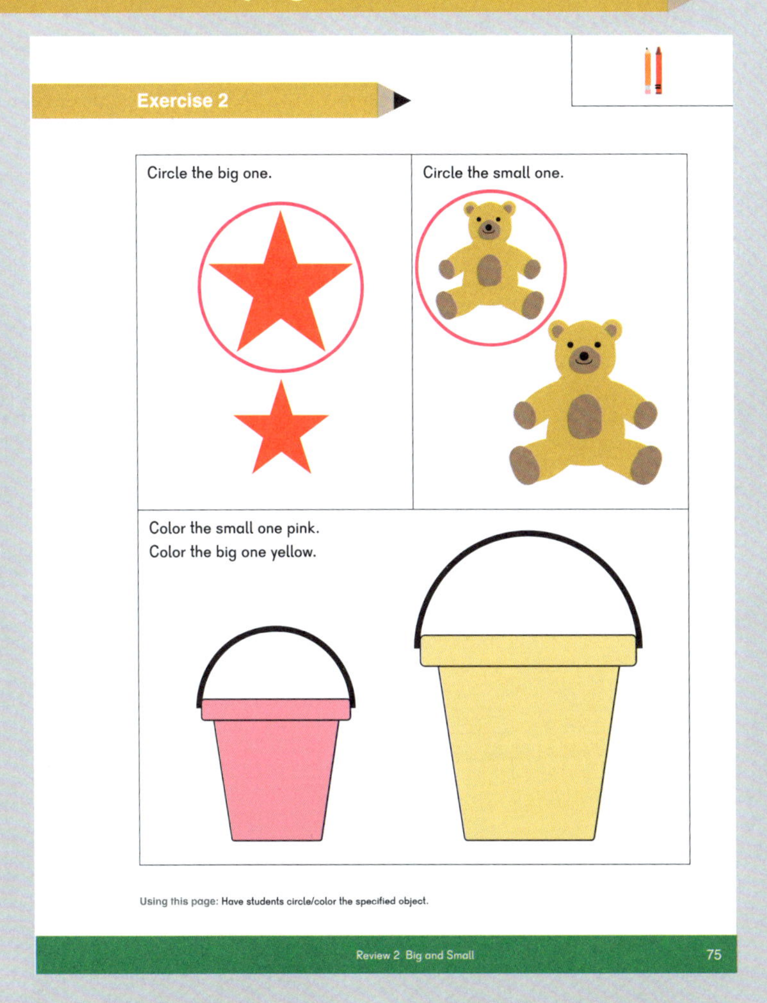

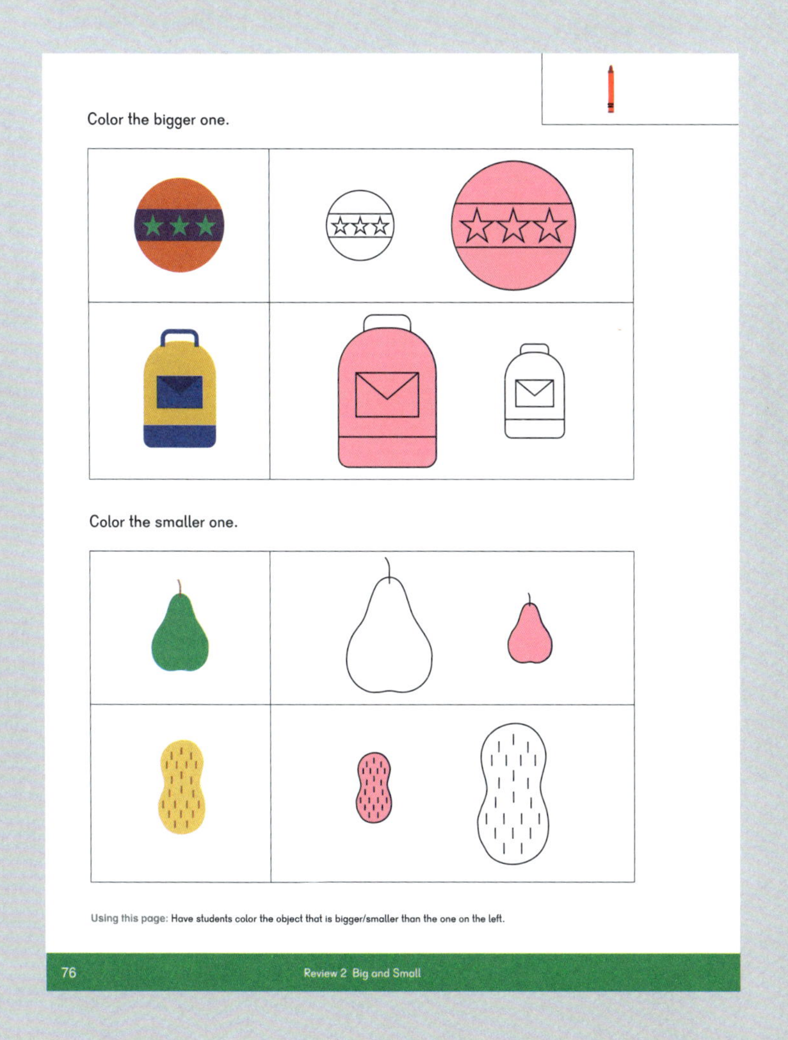

Exercise 3 • pages 77–78

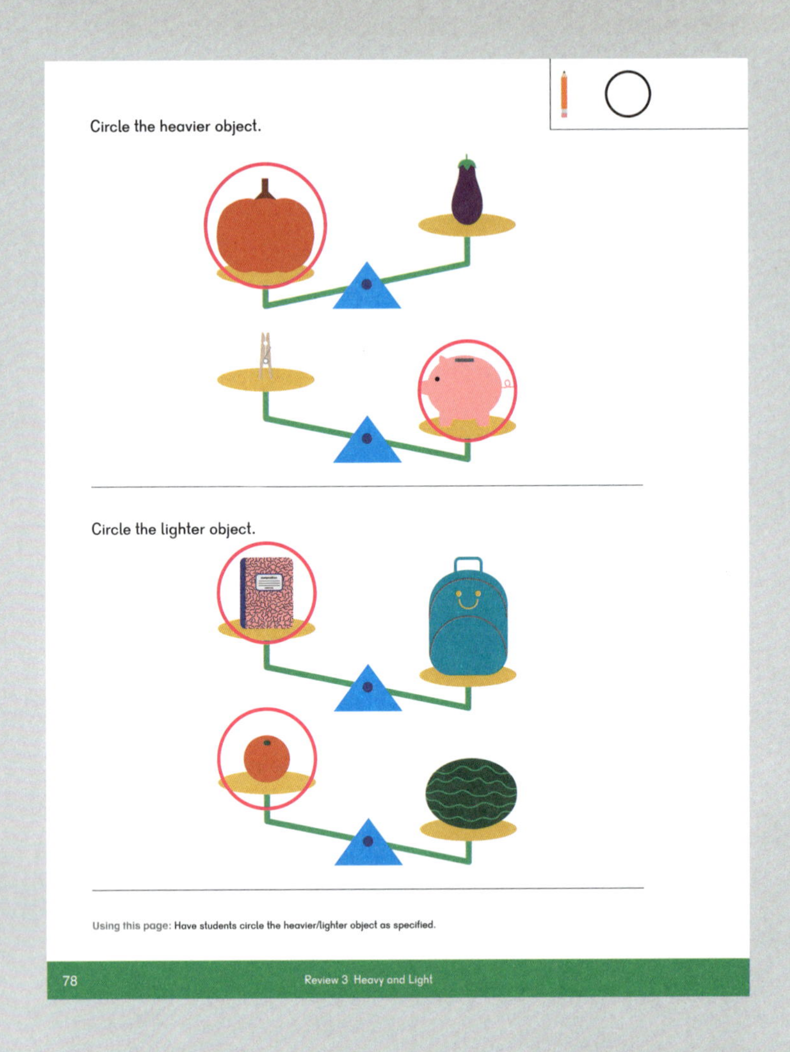

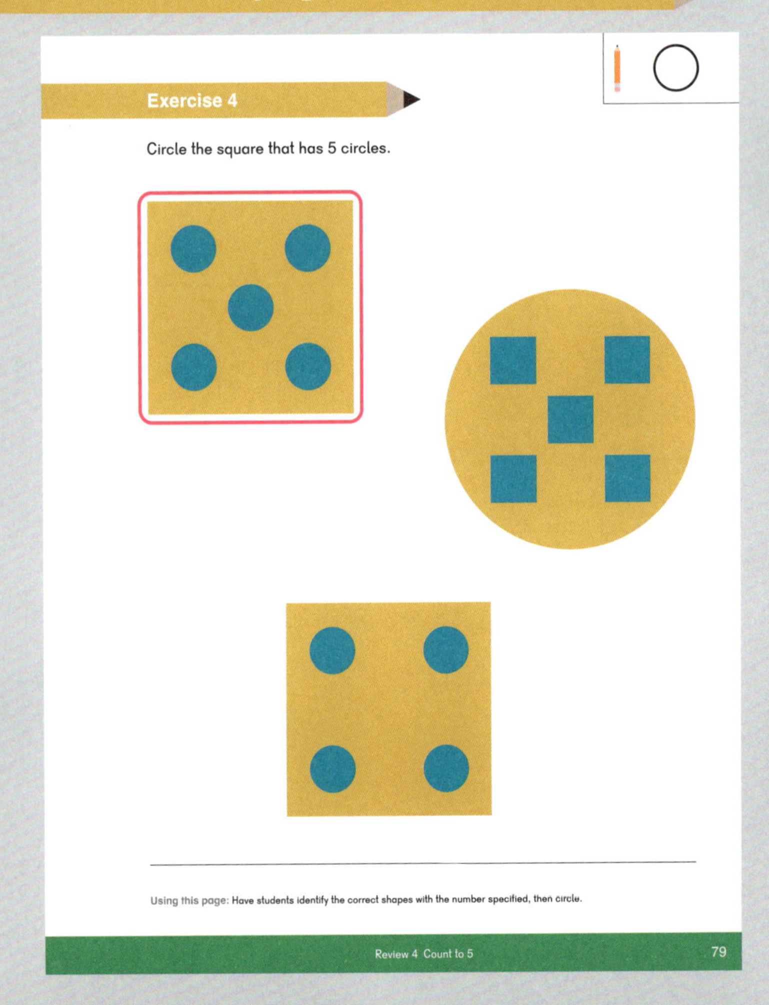

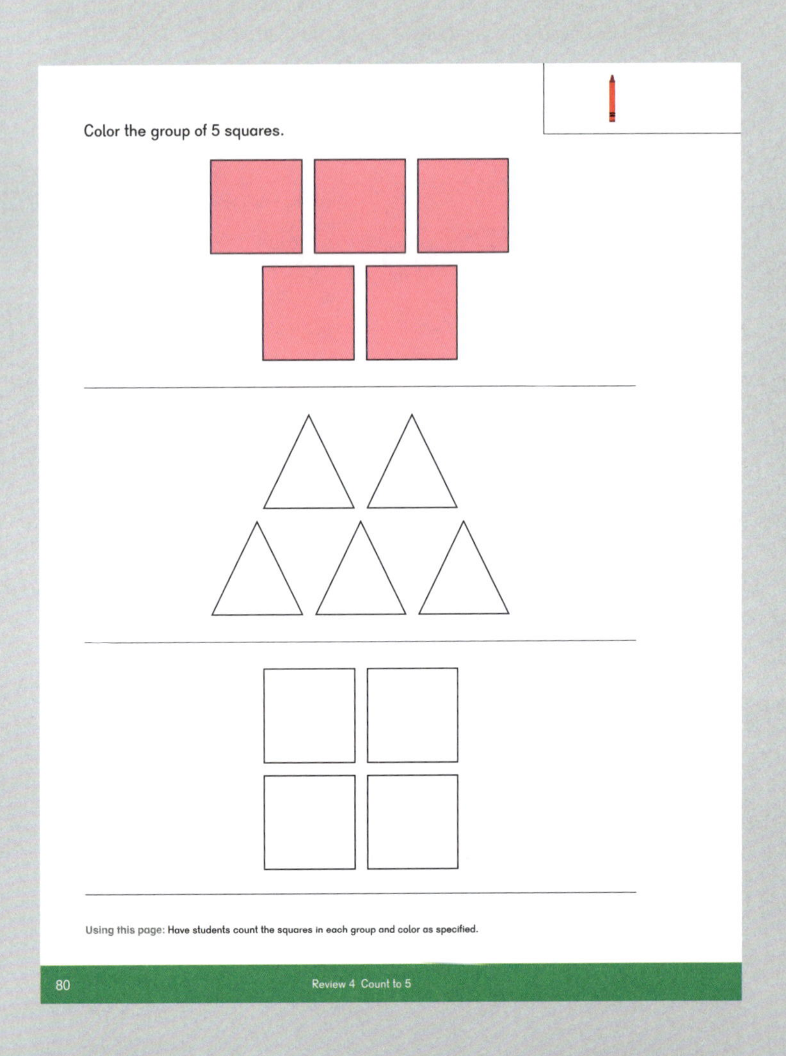

Exercise 5 • pages 81–82

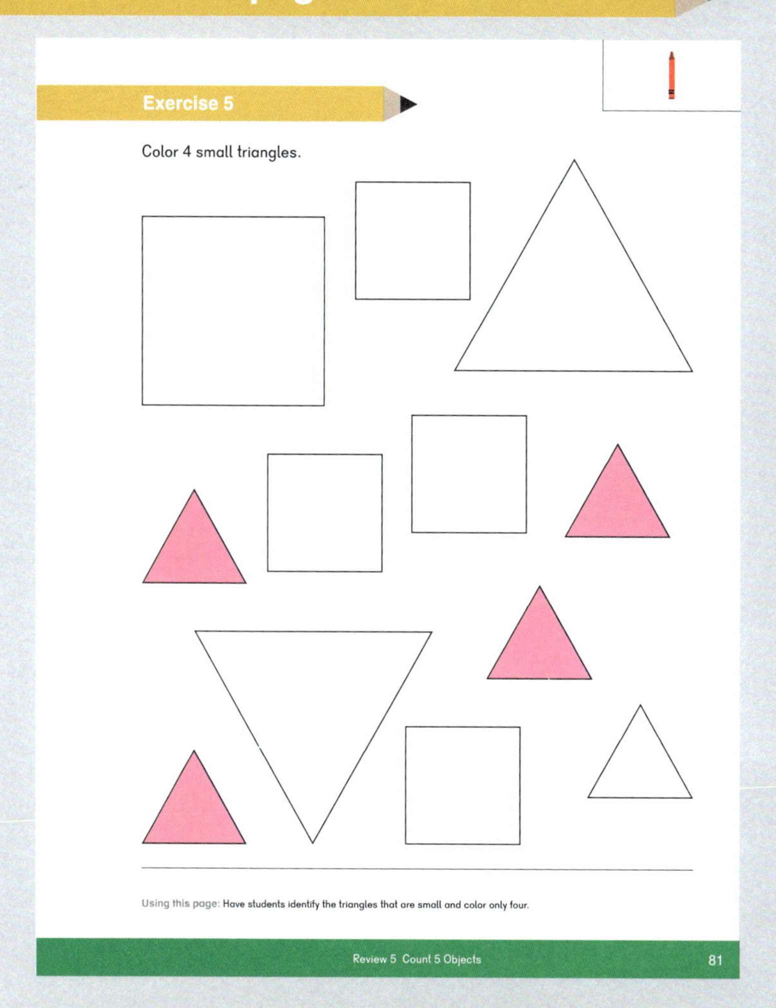

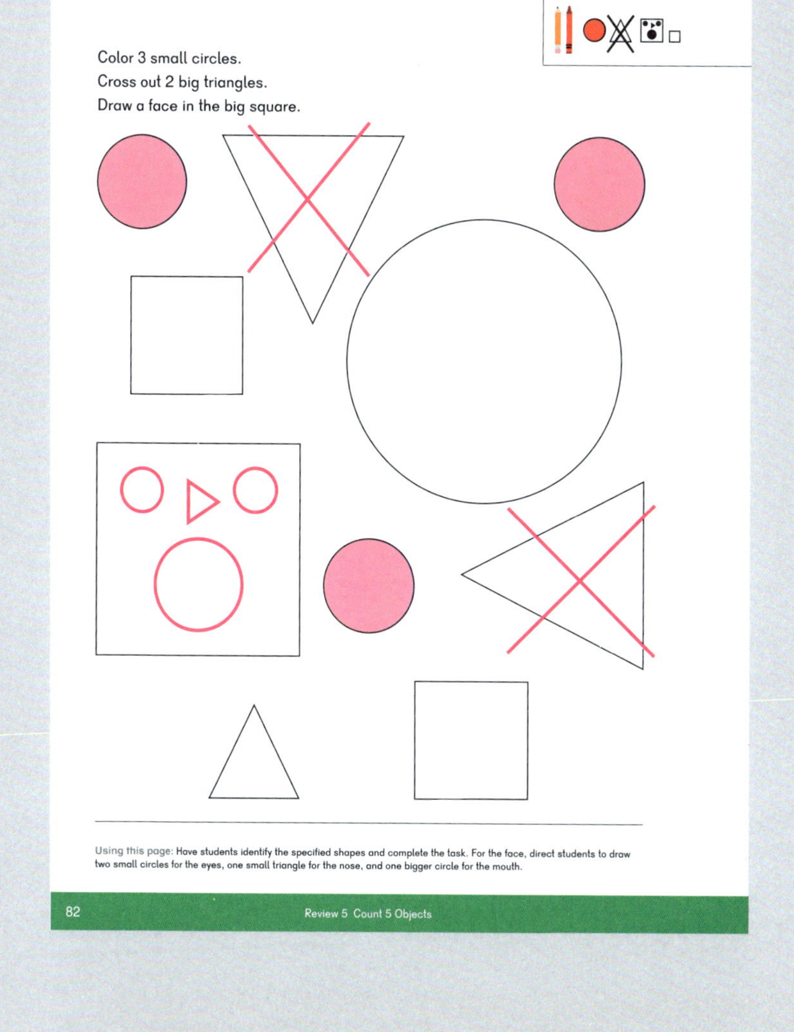

Exercise 6 • pages 83–84

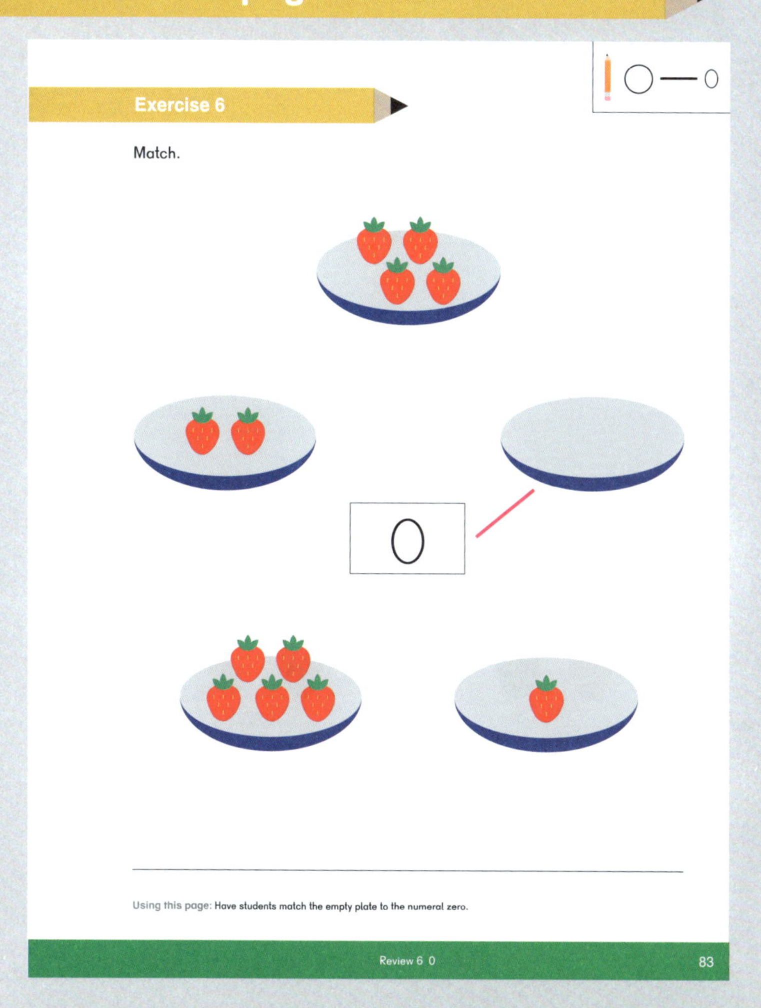

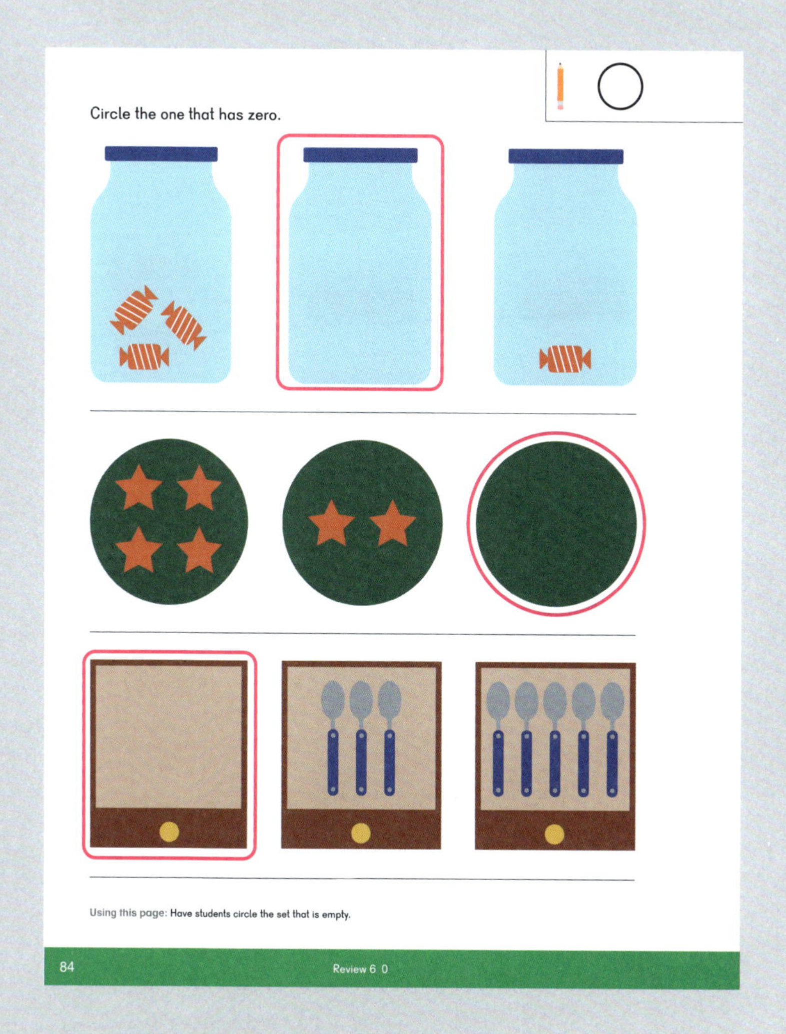

Exercise 7 • pages 85–86

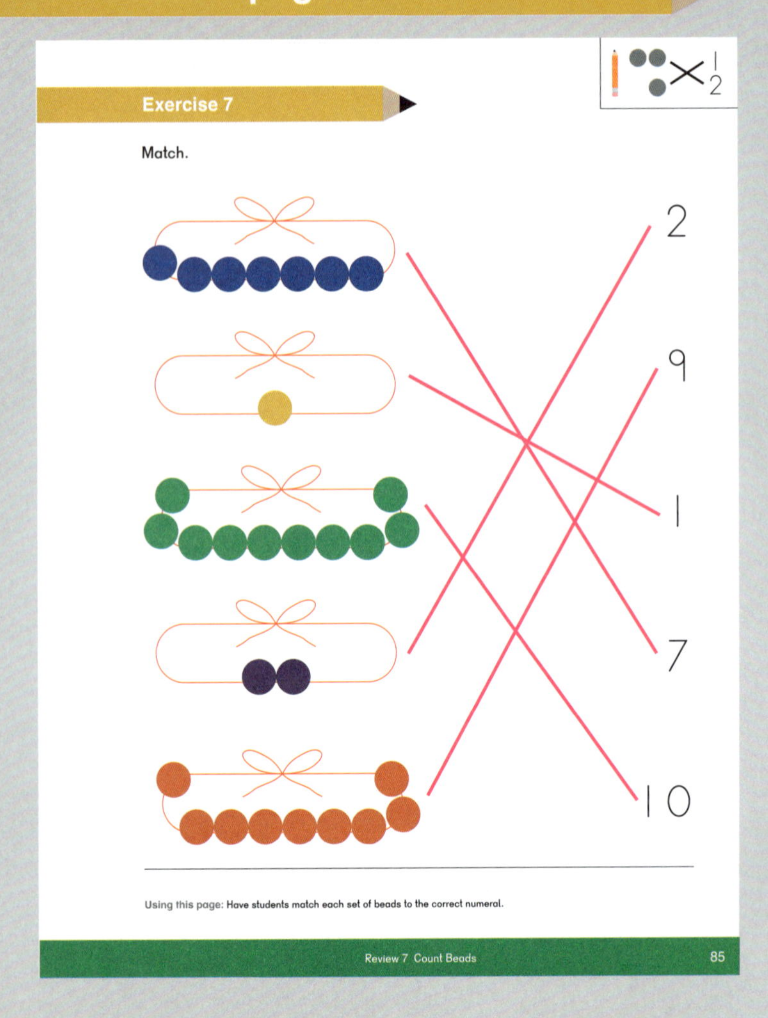

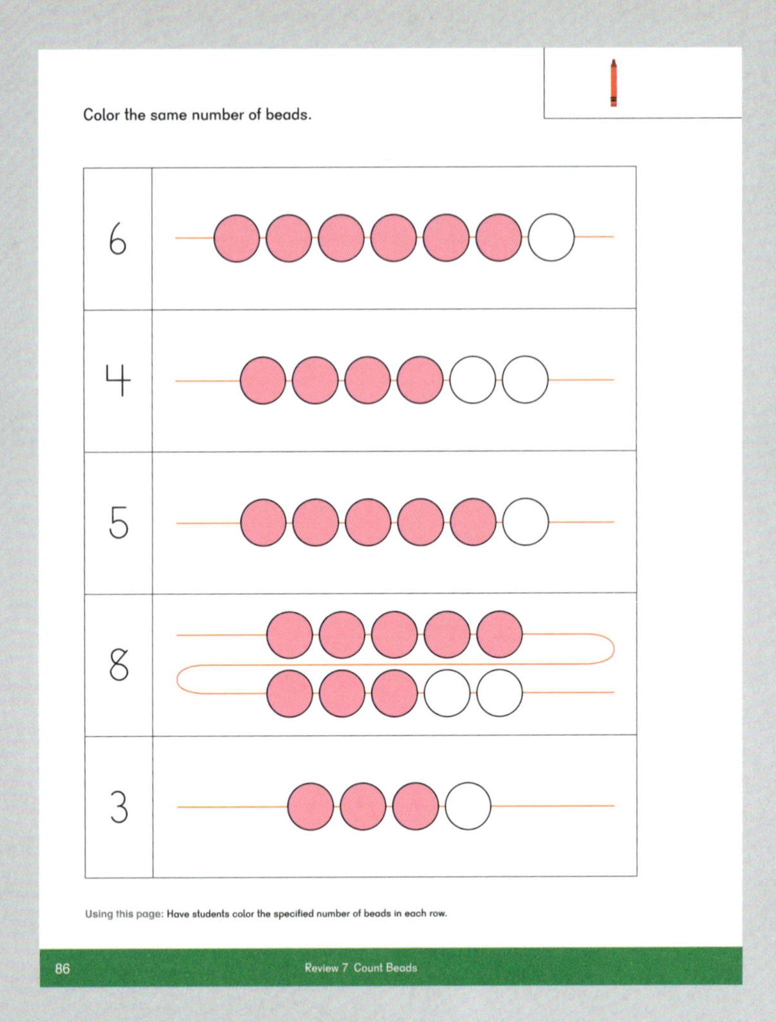

Exercise 8 • pages 87–88

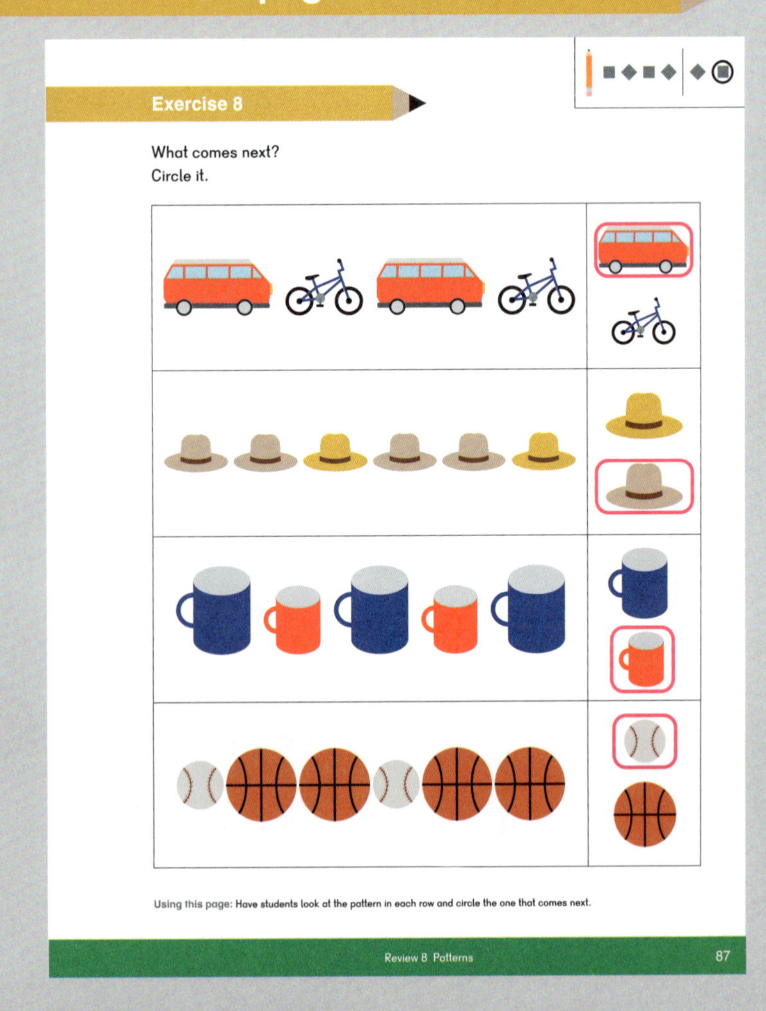

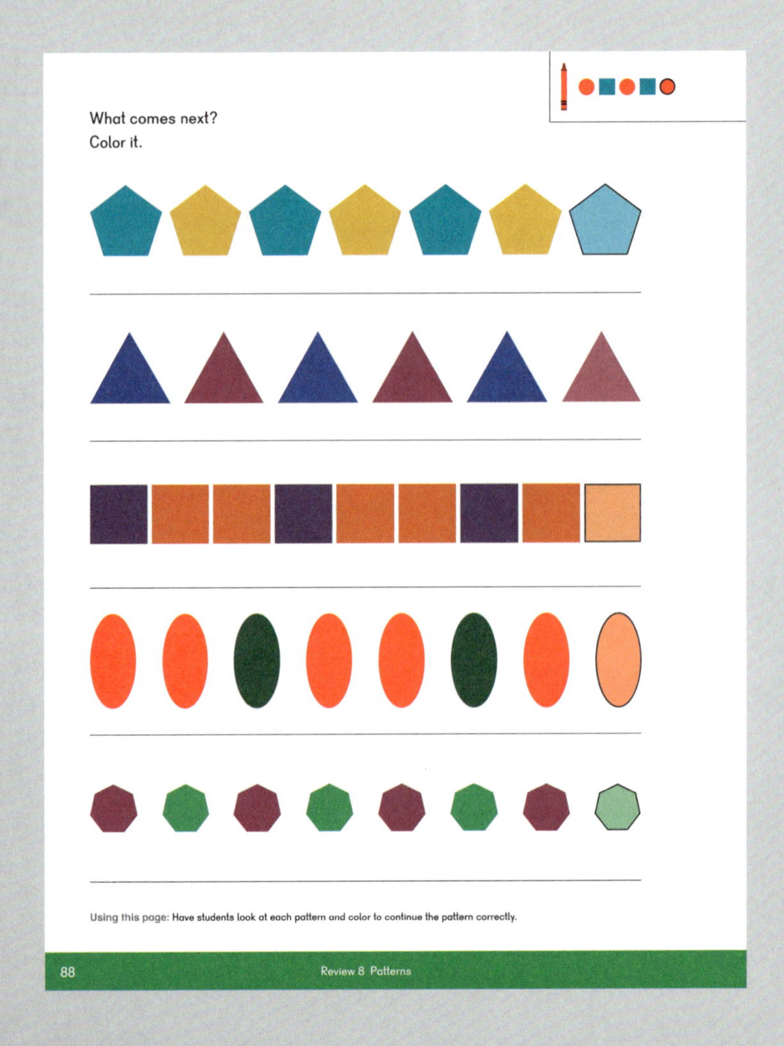

Exercise 9 • pages 89–90

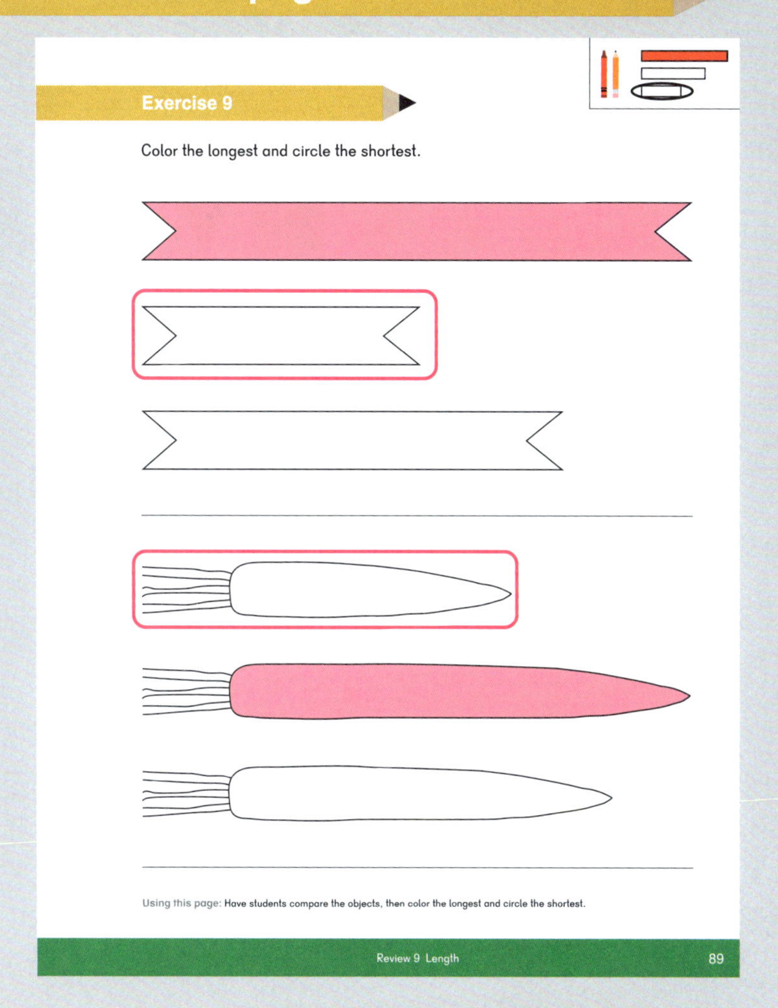

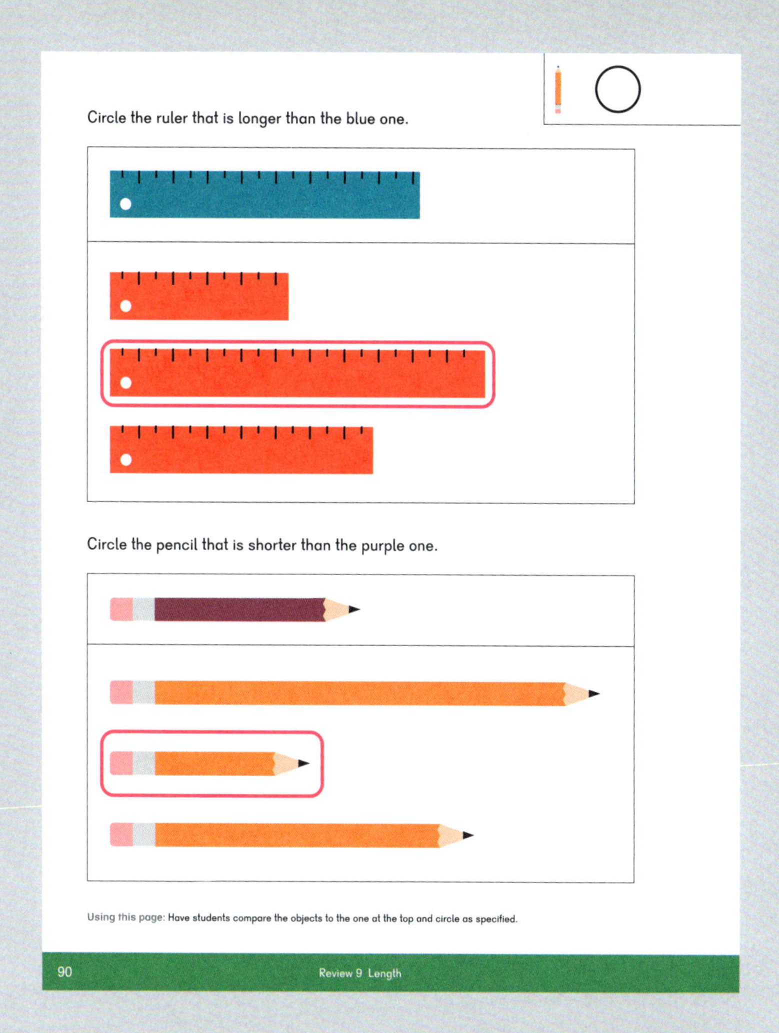

Exercise 10 • pages 91–92

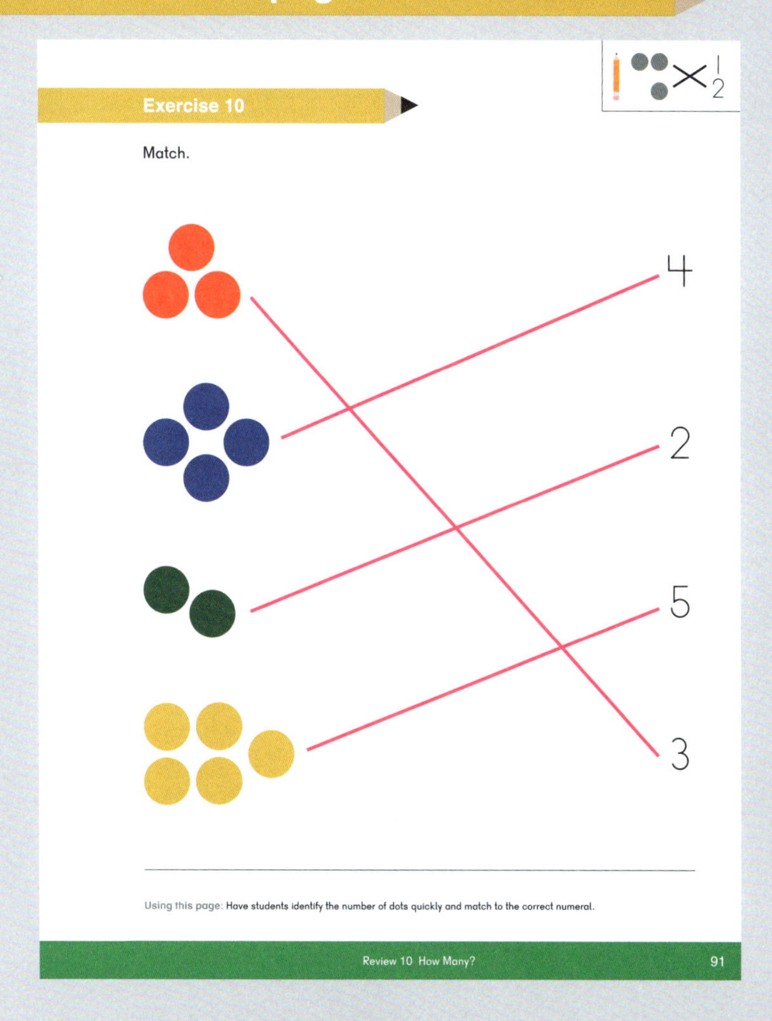

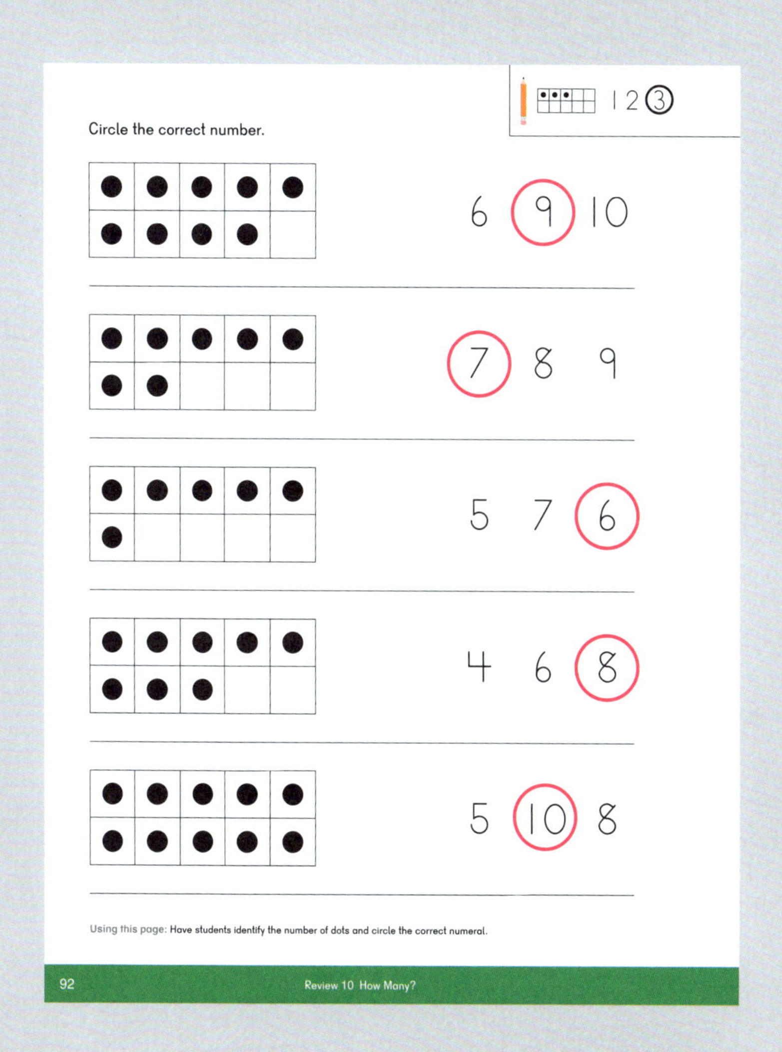

Exercise 11 • pages 93–94

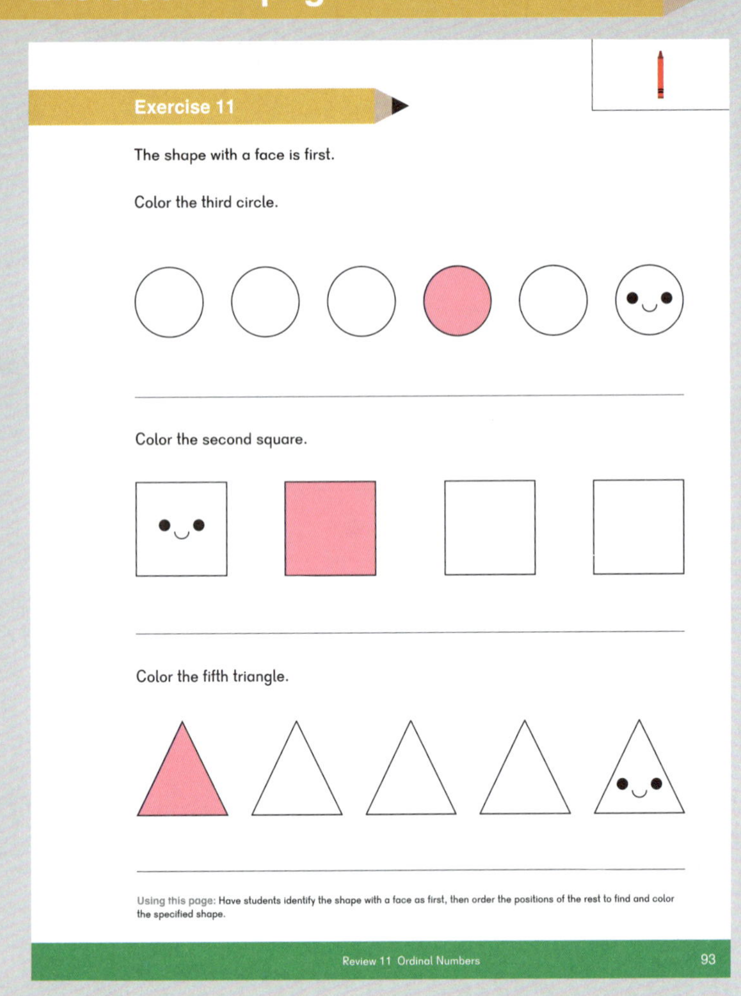

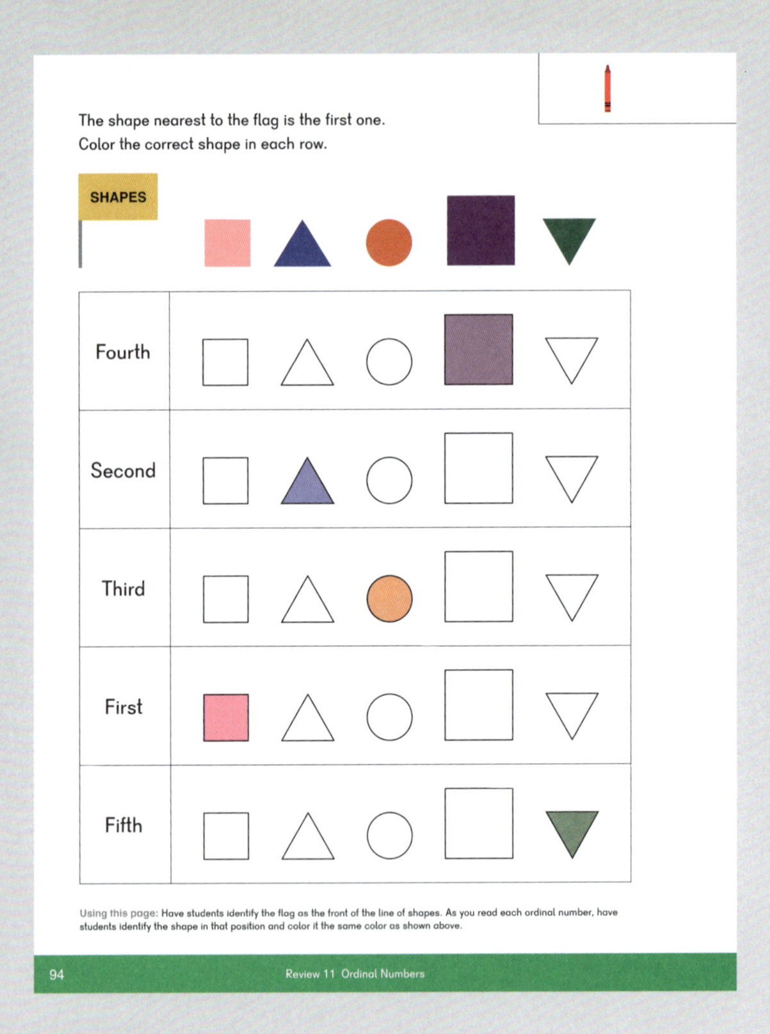

Exercise 12 • pages 95–96

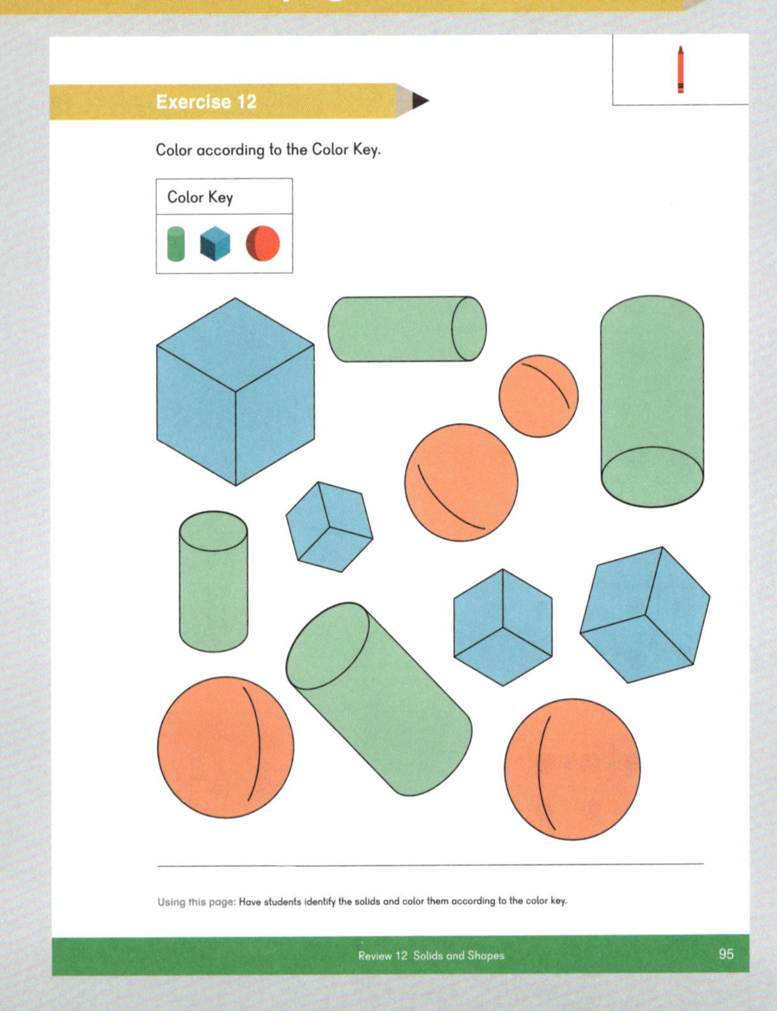

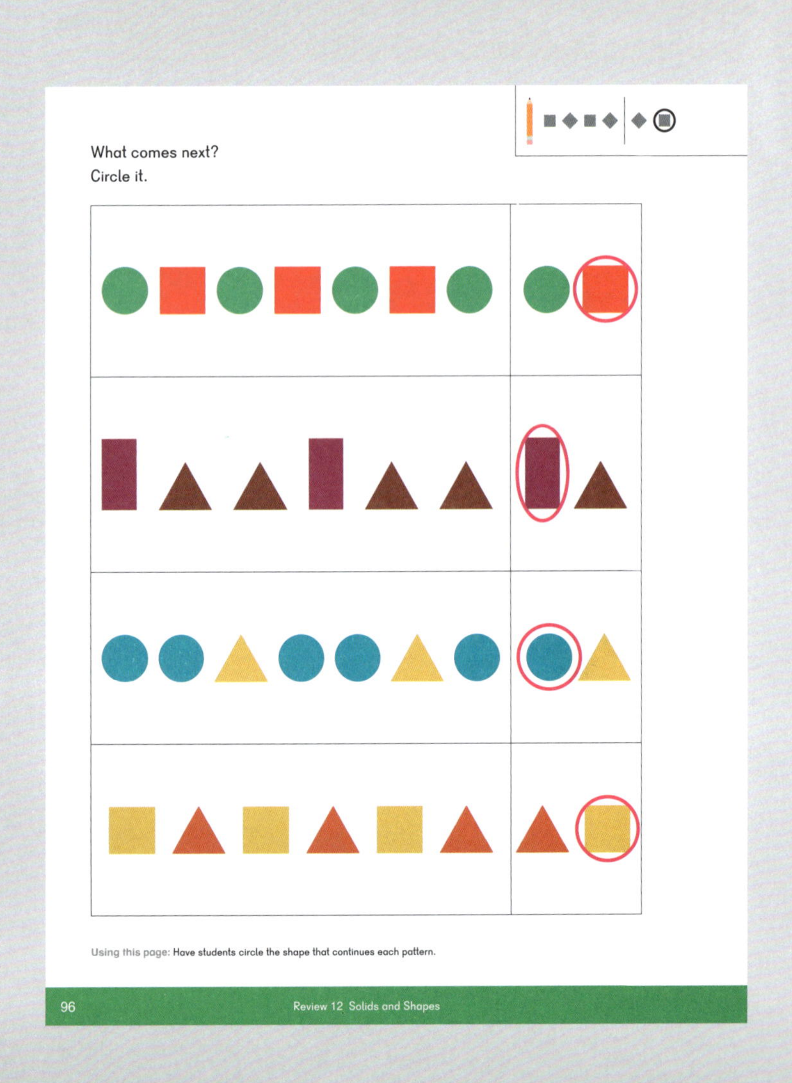

Exercise 13 • pages 97–98

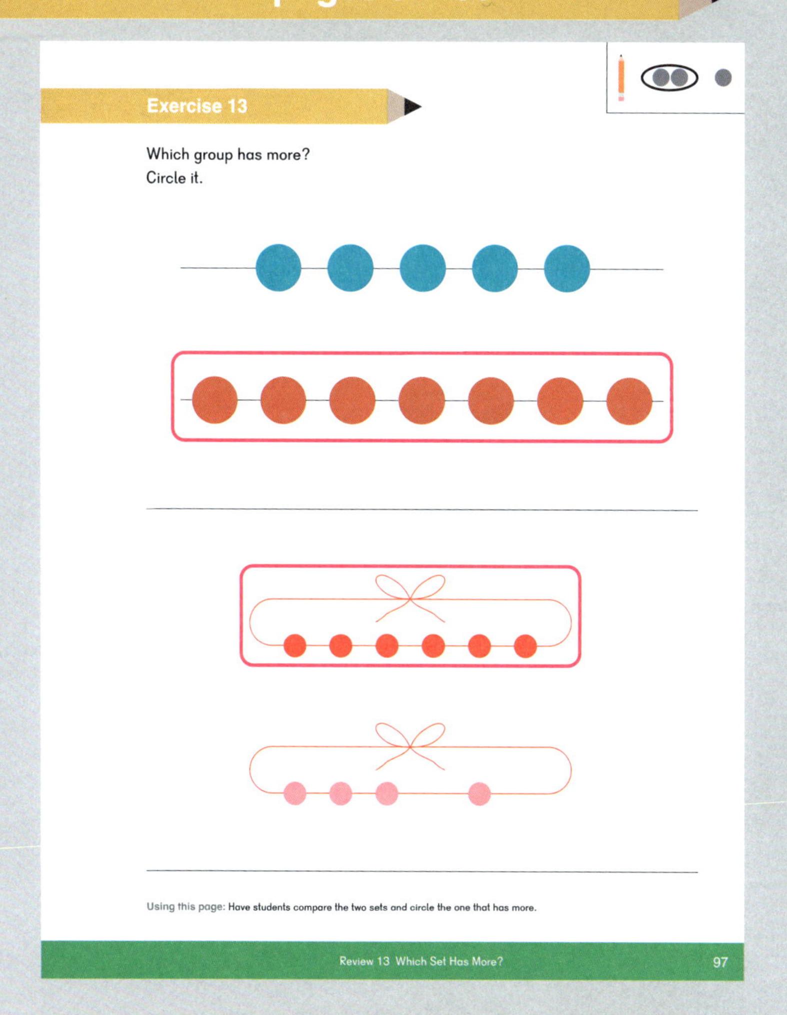

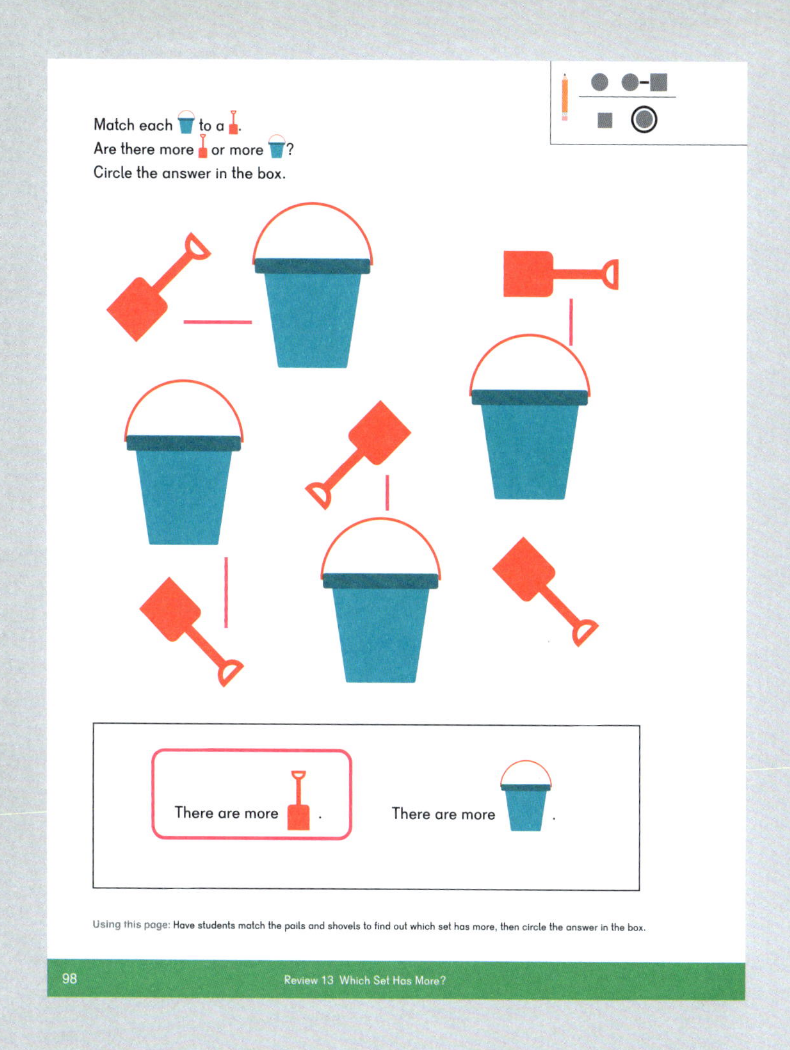

Exercise 14 • pages 99–100

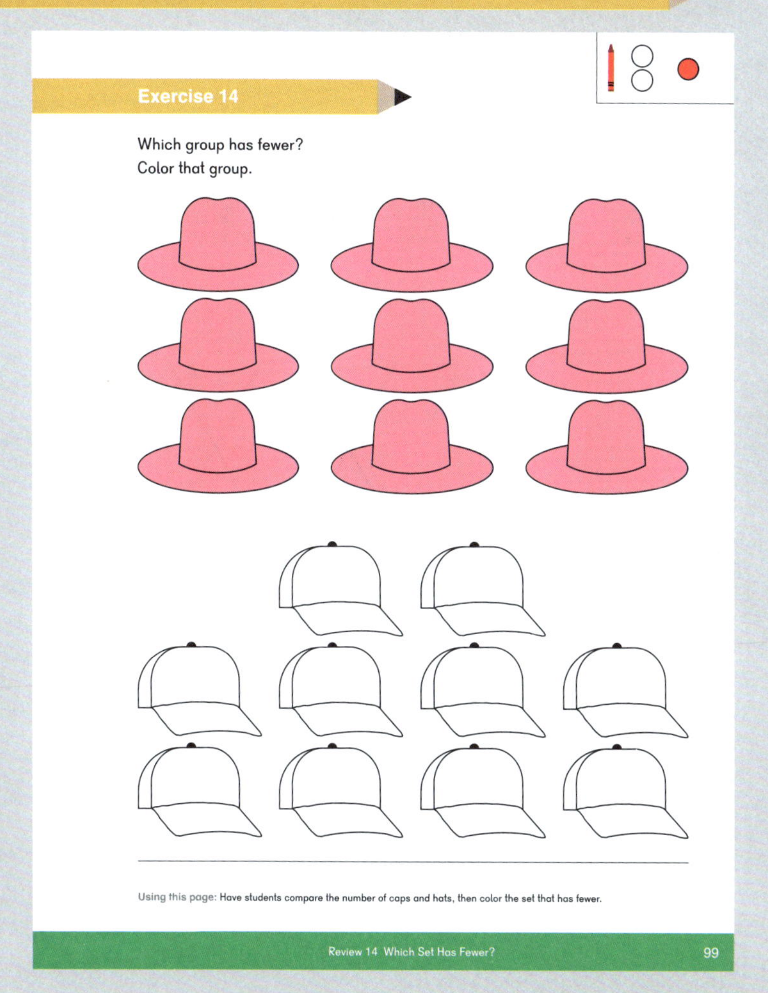

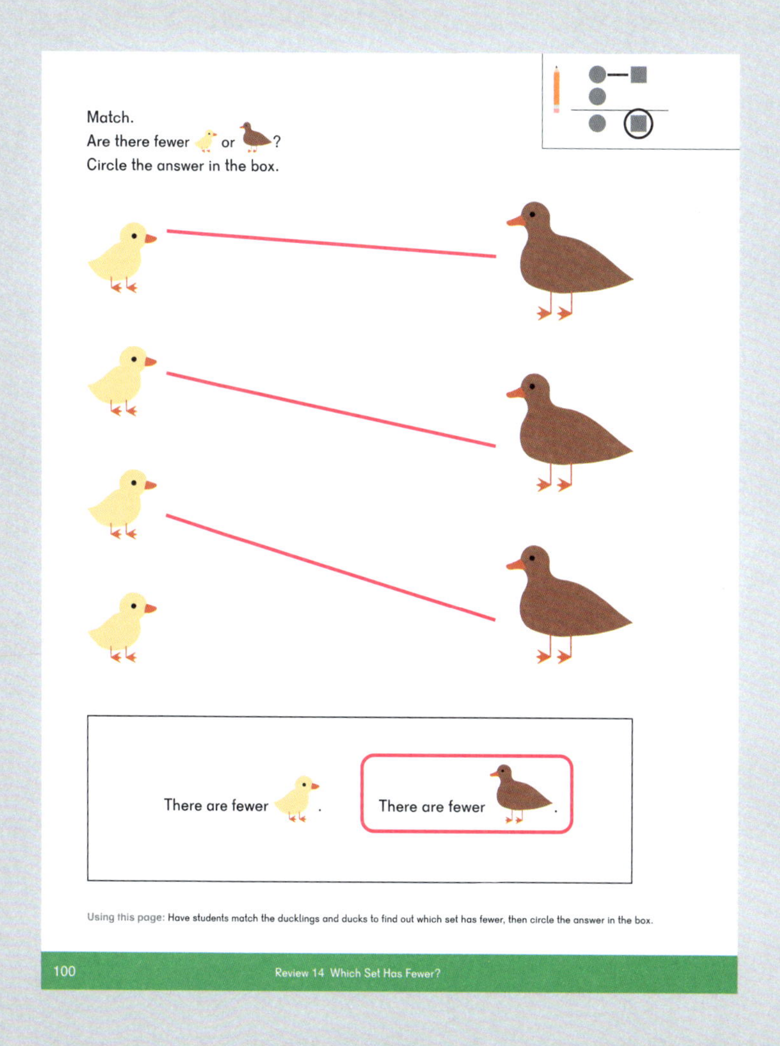

Exercise 15 • pages 101–102

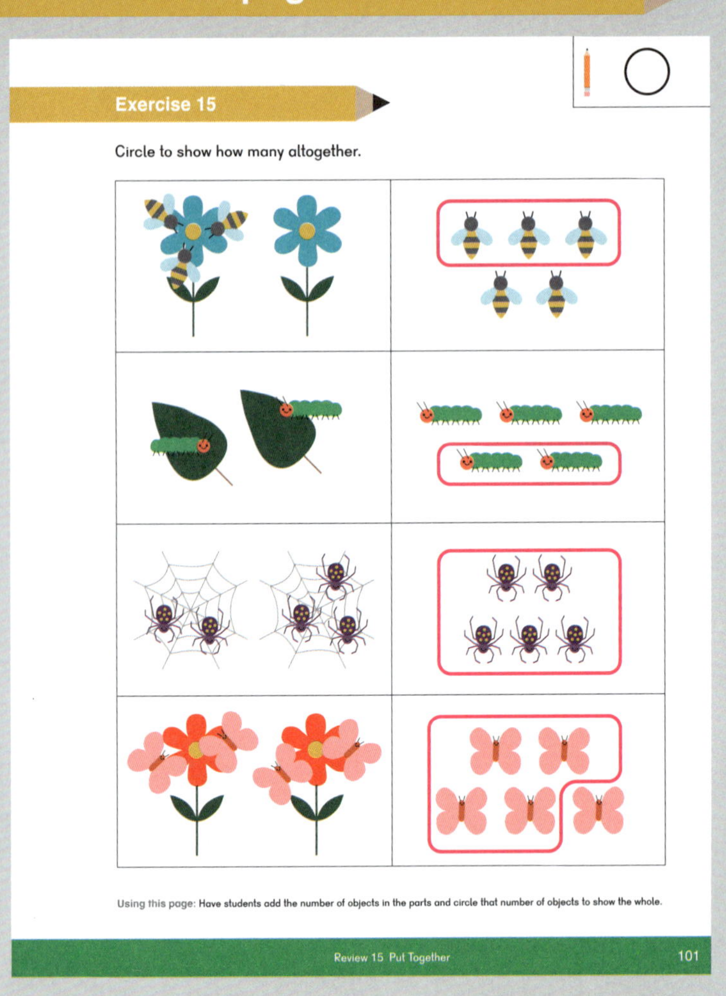

Exercise 15

Circle to show how many altogether.

Using this page: Have students add the number of objects in the parts and circle that number of objects to show the whole.

Review 15 Put Together 101

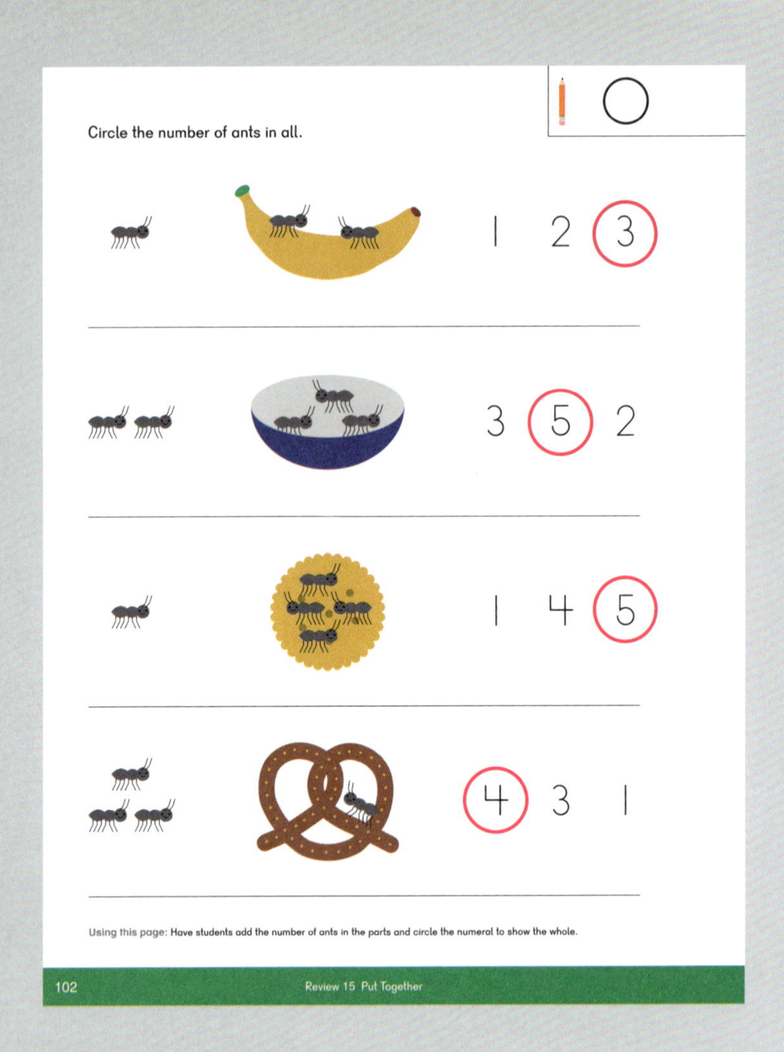

Circle the number of ants in all.

1 2 3

3 5 2

1 4 5

4 3 1

Using this page: Have students add the number of ants in the parts and circle the numeral to show the whole.

102 Review 15 Put Together

Exercise 16 • pages 103–104

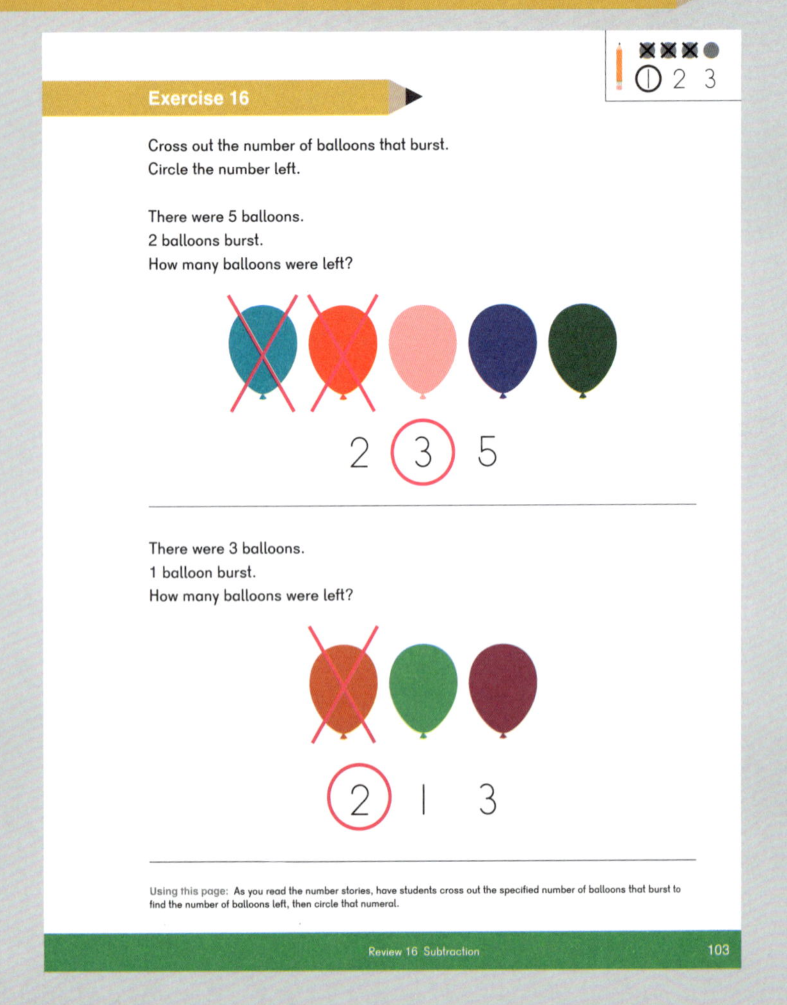

Exercise 16

Cross out the number of balloons that burst.
Circle the number left.

There were 5 balloons.
2 balloons burst.
How many balloons were left?

2 3 5

There were 3 balloons.
1 balloon burst.
How many balloons were left?

2 1 3

Using this page: As you read the number stories, have students cross out the specified number of balloons that burst to find the number of balloons left, then circle that numeral.

Review 16 Subtraction 103

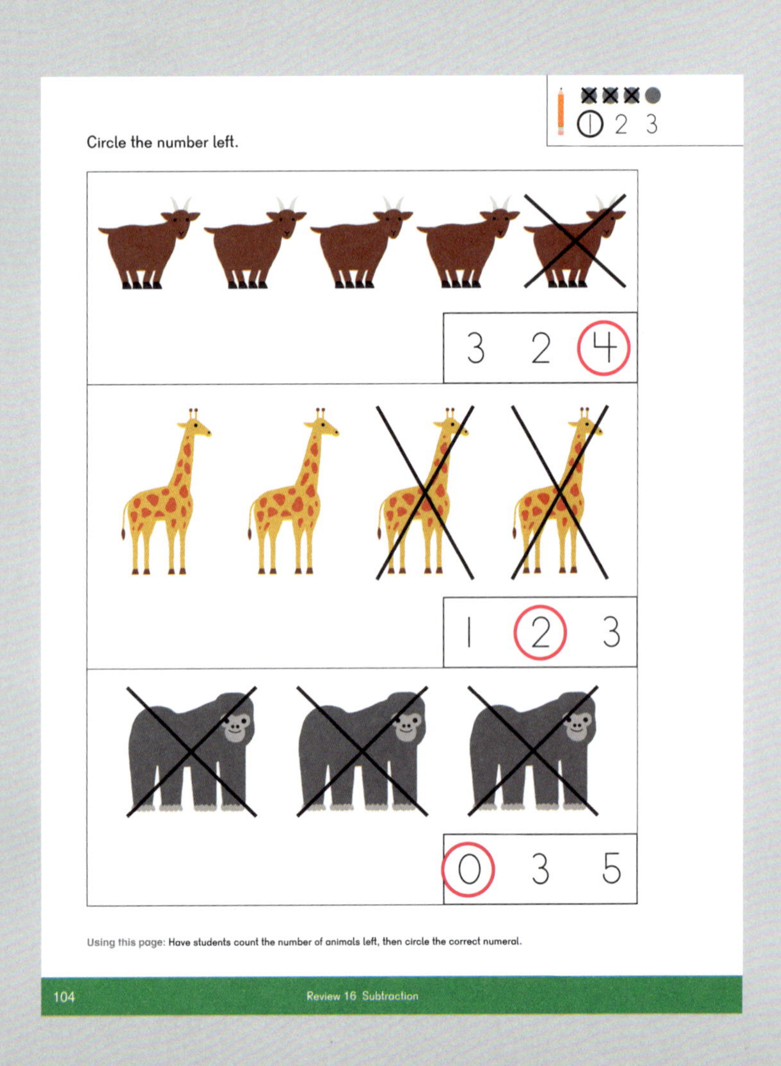

Circle the number left.

3 2 4

1 2 3

0 3 5

Using this page: Have students count the number of animals left, then circle the correct numeral.

104 Review 16 Subtraction

Blackline Masters for PKB

All Blackline Masters used in the guide can be downloaded from dimensionsmath.com.
This lists BLMs used in the **Explore** and **Learn** sections.
BLMs used in **Activities** are included in the Materials lists within each chapter.

0 Art Paper	**Chapter 13:** Review 6
Addition and Equal Symbol Cards	**Chapter 13:** Looking Ahead 4
Addition Facts Cards	**Chapter 13:** Looking Ahead 4
Apple Subtraction Cards	**Chapter 12:** Lesson 6, Lesson 8
Apple Tree Template	**Chapter 12:** Lesson 6
Bicycle/Tricycle Cards	**Chapter 10:** Lesson 3
Blank Five-frames	**Chapter 11:** Lesson 3, Lesson 4, Lesson 5 **Chapter 13:** Review 13
Brownie Bear Template	**Chapter 9:** Lesson 3
Buzzing Bee Template	**Chapter 12:** Lesson 5
Circle Match Cards	**Chapter 13:** Review 5, Review 6
Die	**Chapter 9:** Lesson 2
Domino Cards	**Chapter 11:** Lesson 3, Lesson 6 **Chapter 13:** Review 7, Review 15
Dot Cards	**Chapter 13:** Review 10, Review 14
Five-frame Cards	**Chapter 11:** Lesson 3, Lesson 5 **Chapter 12:** Lesson 7, Lesson 8
Hey Diddle Diddle Cards	**Chapter 10:** Lesson 1
Ladybug Coloring Sheets	**Chapter 11:** Lesson 2
Ladybug Playing Cards	**Chapter 11:** Lesson 2
Linking Cube Template	**Chapter 12:** Lesson 3
Number Cards	**Chapter 11:** Lesson 2, Lesson 3, Lesson 5, Lesson 6 **Chapter 12:** Lesson 2, Lesson 3, Lesson 5, Lesson 6, Lesson 8 **Chapter 13:** Review 5, Review 6, Review 7, Review 10, Review 15, Review 16, Looking Ahead 4, Looking Ahead 5, Looking Ahead 7
Number Cards — Large	**Chapter 11:** Lesson 2 **Chapter 13:** Review 7, Looking Ahead 4, Looking Ahead 7

Paper Cutouts — Chapter Opener	**Chapter 9:** Chapter Opener, Lesson 10
Paper Cutouts — Lesson 5	**Chapter 9:** Lesson 5, Lesson 10
Paper Cutouts — Lesson 7	**Chapter 9:** Lesson 7, Lesson 10
Paper Cutouts — Lesson 8	**Chapter 9:** Lesson 8, Lesson 10
Paper Cutouts — Lesson 9	**Chapter 9:** Lesson 9, Lesson 10
Paper Cutouts — Review 4	**Chapter 13:** Review 4
Parts and Whole Cards	**Chapter 12:** Lesson 1, Lesson 2, Lesson 8
Parts and Whole Mats	**Chapter 11:** Lesson 6 **Chapter 13:** Review 15, Looking Ahead 4, Looking Ahead 5
Parts Mat Template	**Chapter 12:** Lesson 2
Picture Cards	**Chapter 13:** Looking Ahead 7
Picture Ten-frame Cards	**Chapter 10:** Lesson 2
Square-Triangle Flash Cards	**Chapter 13:** Review 4
Subtraction and Equal Symbol Cards	**Chapter 13:** Looking Ahead 5
Ten-frame Cards	**Chapter 10:** Lesson 2, Lesson 4 **Chapter 13:** Review 10
What's Left? Cards	**Chapter 12:** Lesson 5
Word Cards	**Chapter 13:** Looking Ahead 7
Word-Picture Cards	**Chapter 13:** Looking Ahead 7